COMET

COMET

Carl Sagan and Ann Druyan

Michael Joseph
London

First published in Great Britain by Michael Joseph Ltd
44 Bedford Square, London WC1
1985

British Library Cataloguing in Publication Data

Sagan, Carl
Comet.
1. Comets
I. Title II. Druyan, Ann
523.6   QB721

ISBN 0 7181 2631 9

Printed in Italy by A. Mondadori Editore - Verona

Title page: *A comet streaks into the Jovian atmosphere.*
(Painting by Don Dixon.)
Part One: *The early Earth at the time of the origin of life.*
(Painting by Kazuaki Iwasaki.)
Part Two: *A dinosaur contemplates its fate during the*
postulated late Cretaceous comet shower.
(Painting by Jon Lomberg.)
Part Three: *A comet over Vallis Marineris, Mars.*
(Painting by Kim Poor.)

CONTENTS

To Shirley Arden
for more than a decade of friendship
and hard work well done
with our love and admiration

INTRODUCTION

Before the Earth was formed, there were comets here. Afterwards, and for all subsequent eons, comets have graced our skies. But until very recently, the comets performed without an audience; there was, as yet, no consciousness to wonder at their beauty. This all changed a few million years ago, but it was not until the last ten millennia or so that we began to make permanent records of our thoughts and feelings. Ever since, comets have left a good deal more than dust and gas in their wakes; they have trailed images, poetry, questions, and insights. In this book, we have sought to rediscover those trails, to explore our present understanding of the comets, and to speculate on what else may be possible. We were inspired by the 1985/86 return of one of the Earth's most brilliant (and punctual) visitors, Halley's Comet.

We have tried to give some sense of the evolution of scientific discovery, and present the evidence for and against cometary theories that have been rejected, those that are currently fashionable, and some that are neither fashionable nor rejected. We hope we have indicated clearly which is which. Certain concepts and terms are presented more than once for clarity and accessibility by the general reader. This is not by any means a mathematical book, but all modern science is quantitative and so we have included a little arithmetic here and there. When we want to know how far away something is, we could measure in miles or versts or leagues. Nature is just the same no matter what measuring rods we use. But because it is the scientific convention, because the vast majority of nations on the planet have adopted it, and because it is simpler, we will here employ the metric system. A metre is about a yard long. A thousand of them make a kilometre, which is about sixty percent of a mile. If we imagine a metre divided into a hundred equal parts, each part is a centimetre. Around two-and-a-half centimetres make an inch. A micron is a millionth of a metre, much too small to see. Ten thousand atoms, shoulder-to-shoulder, make a chain one micron long. Your fingernail is about ten thousand microns wide, around one centimetre.

For reasons of length and readability, we have included the names of only a tiny fraction of the cometary scientists alive today – far more of them exist than the total number of cometary scientists who have lived in all previous generations – and who have done so much to transform the subject into one of the most exciting areas of modern science. We beg the

1 *The orbit of Halley's Comet, shown in yellow. The blue circle represents the Earth's orbit around the Sun. Halley's Comet spends most of its time in the outer solar system, sweeping inside the orbit of the Earth once every seventy-six years. It was last at its farthest point from the Sun (aphelion), in 1948; its most recent passage close to the Sun (perihelion) was in 1985/86. The comet moves very slowly when it is far from the Sun and much more quickly when it is close to the Sun. (Diagram by Jon Lomberg/BPS.)*

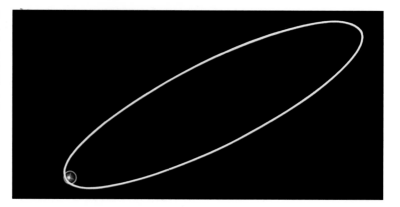

indulgence of those whose names we have omitted. The interested reader can follow the track of original papers to find the names of these specialists through the bibliography in the back of the book. In that bibliography, we have included both popular books on astronomy and representative or especially interesting works from the scientific literature.

We are indebted to the world astronomical community for their generosity to us throughout the preparation of this book. Joseph Veverka of Cornell University, a moving force behind US plans for spacecraft missions to the asteroids and the comets, served as chief technical consultant and reviewer, and provided a large number of the astronomical photographs in the book. We are grateful to Mary Roth for facilitating the acquisition of these photographs.

Several colleagues were kind enough to read earlier drafts of this book and provide valuable comments: Martha Hanner, Joseph Marcus, Steven Soter, Paul Weissman and Donald Yeomans, who also calculated the dates of future returns of Halley's Comet. Others who permitted us to tap their expertise include Donald Brownlee, Brian Marsden, Marcia Neugebauer, Ray Newburn, Zdenak Sekanina, Reid Thompson, and Fred Whipple.

One of the most pleasant experiences in writing this book was our encounter with Ruth S. Freitag, Senior Science Specialist of the Science and Technology Division of the Library of Congress. Ms Freitag has recently published a bibliography of Halley's Comet with over 3,200 citations. The subject will deserve a revised edition soon. Ms Freitag's knowledge, enthusiasm for her subject, and willingness to share a treasury of comet illustrations speaks well of the Nation's Library.

The visual content of *Comet* is largely due to Jon Lomberg, an artist whose dreams are informed by science. His diagrams – produced with Simon Bell and Jason LeBel of Bell Production Services, Toronto, Canada – teach painlessly and with elegance. In addition to reviewing the manuscript, Jon also coordinated the efforts of ten artists in producing some forty specially commissioned paintings for this book. We are proud to display herein the work of some of the planet's finest astronomical artists: Michael Carrol, Don Davis, Don Dixon, William K. Hartmann, Kazuaki Iwasaki, Pamela Lee, Jon Lomberg, Anne Norcia, Kim Poor and Rick Sternbach.

The manuscript was guided through its many incarnations by Shirley Arden, whom we acknowledge in the dedication. We also thank Pandora Peabody and Maruja Farge, who in different ways made fundamental contributions to this book.

Patricia Parker showed us the possibilities of a concordance of literary references to comets, launching us on a heady voyage. We have benefitted from the remarkable breadth and depth of the scholarly community at Cornell University. Patricia Gill provided us with original tranlations of numerous ancient comet references and assembled a corps of talented researchers and translators.

We thank Howard Kaminsky, Jason Epstein, Laura Schultz and Bob Aulicino at Random House for their efficient production; Derek Johns for helpful copy-editing; Scott Meredith, Jack Scovil, Jonathan Silverman, and others at the Scott Meredith Literary Agency for services that go beyond the usual norms in making *Comet* possible. We are also grateful to many others, many of whom are listed in the acknowledgement page of the book.

This book, and our knowledge of the solar system, owes much to the openness and relative freedom from secrecy of the National Aeronautics and Space Administration of the United States of America and its counterpart institutes in the Academy of Sciences of the Union of Soviet Socialist Republics.

The thirtieth recorded apparition of Halley's Comet brings us face to face with the great question of our time. This return is the first since we have become a spacefaring civilization, and the first since we have devised the means to destroy ourselves. We recall those eons during which there were no beings on Earth to marvel at the comets; we hope, at least until the Sun dies, that will never be true again.

You lift your children above your shoulders that they may better see a comet and, in so doing, join a chain of generations that stretches back far beyond the reach of written memory. There is nothing more important than protecting that ancient and most precious continuity.

Carl Sagan and Ann Druyan
Ithaca, New York
6 August 1985

PART ONE
THE NATURE
OF COMETS

Chapter 1

ASTRIDE THE COMET

How vast is creation! I see the planets rise and the stars hurry by, carried along with their light! What, then, is this hand which propels them? The sky broadens the more I ascend. Worlds revolve around me. And I am the centre of this restless creation.

Oh, how great is my spirit! I feel superior to that miserable world lost in the immeasurable distance beneath me; planets frolic about me – comets pass by casting forth their fiery tails, and centuries hence they will return, still running like horses on the field of space. How I am soothed by this immensity! Yes, this is indeed made for me; the infinite surrounds me on all sides. I am devouring it with ease.

Gustave Flaubert,
Smarh (1839,
translated by James LeBlanc).

These are the snows of yesteryear, the pristine remnants of the origin of the solar system, waiting frozen in the interstellar dark. Here, trillions of orbiting snowbanks and icebergs are stored, gently suspended about the Sun. They cruise no faster than a small propeller-driven aircraft would, buzzing through the blue skies of far-off Earth. The slowness of their motion just balances the gravity of the distant Sun, and poised between feeble contending forces, they take millions of years to complete one orbit around that yellow point of light. Out here, you are a third of the way to the nearest star. Or rather, to the next nearest star: in the depth and utter blackness of the dark sky around you, it is entirely clear that the Sun is one of the stars. It is not even the brightest star in the sky. Sirius is brighter, and Canopus. If there are planets circling the star called the Sun, there is no hint of them from this remote vantage point.

These trillions of floating icebergs fill an immense volume of space; the nearest one is three billion kilometres away from you, about the distance of the Earth from Uranus. There are many icebergs, but the space they fill, a thick shell surrounding the Sun, is incomprehensibly vast. Most of them have been out here since the solar system began, quarantined from whatever mischief may be going on down there, in that alien and hostile region bordering the Sun.

Beyond the occasional soft ping of a cosmic ray from some collapsed star at the other end of the Milky Way, hardly

1.1 *(Opposite) The Earth as seen from the icy surface of a comet – clearly a member of a shower of comets that has entered the inner solar system some time in the remote past. (Painting by Jon Lomberg.)*

anything ever happens here. It is very peaceful. But something *has* happened, a gravitational intrusion, not by the Sun or its possible planets, but by another star. It was slow in coming, and at its closest, it was never very near. You can see it over there, glowing faintly red, much dimmer than the Sun. This cloud of icebergs has been carried with the Sun on its motions through the Milky Way galaxy. But other stars have their own characteristic motions, and sometimes by accident approach us. So every now and then, as now, there is a little gravitational rumbling, and the cloud trembles.

Since your iceberg is bound so weakly to the Sun, even a little push or tug is enough to throw it onto some new trajectory. The neighbouring icebergs – much too small and distant for you to see directly – have been similarly affected, and are now hurrying off in many directions. Some have been shaken loose from the gravitational shackles that had bound them to the Sun, and are now liberated from their ancient servitude, embarking on odysseys into the vast spaces between the stars. But for your iceberg, there is a different destiny working itself out: you have been tugged in such a way that you are now falling, slowly at first, but with gradually increasing speed – down, down, down towards the point of light about which this vast collection of little worlds slowly revolves.

Imagine that you are as patient and as longlived as the iceberg on which you are standing, that you have adequate life expectancy and life support equipment for a journey of a few million years. You are falling towards the bright yellow star. Your worldlet and its brethren have been given a name. Comets, they are called. Your comet is an emissary from the kingdom of ice to the infernal realm near the Sun.

Out here, a comet is only an iceberg. Later on, the iceberg will be just one part of the comet, called the nucleus. A typical cometary nucleus is around a kilometre across. Its surface area is the size of a small city. If you were standing on it you would see the smoothly curving contours of gracefully sculpted hillocks built of dark, reddish-brown ice. There is no air on this small world, nothing liquid, and – apart from you, yourself – nothing alive, at least so far as you can see. You can, over the following millions of years, explore every corner, every mountain, every crevice. With the skies perfectly clear, and with no particularly urgent tasks before you, you can also spend a little time studying the magnificent array of unwinking bright stars that surrounds you.

Your footprints are very deep, because the snow beneath your feet is weak. In a few places there are patches of ground so fragile that were you to walk upon them imprudently you would fall through – as in the legendary quicksand of Earth –

onto deep shadowed ice, perhaps metres below. Your fall would be slow, though, almost languorous, because the acceleration caused by gravity, the downward force you feel here on the comet, is only a few thousandths of a percent of the familiar 1 g of Earth.

On more solid ground, the low gravity might tempt you into unprecedented athletic feats. But you must be careful. If you so much as stride purposefully, you walk off the comet altogether. With only a little effort, from a standing position you could leap thirty kilometres into space, taking almost a week to reach the peak of your trajectory. There, gently tumbling, you would have a comprehensive view of the comet slowly rotating beneath you. You can make out its lumpy shape; it is far from a perfect sphere. Perhaps you worry that you have jumped too high, that you will not fall back to the comet, that you will drift alone through space forever. But no, you see that your outbound velocity is gradually diminishing and eventually, ten days after making this modest exertion, you tumble lightly back onto the sombre snows. On this world, you are dangerously strong.

Since it is hard to take a step without launching yourself on a small parabolic trajectory, team sports would be played in agonizing slow motion, the cluster of players rising and spinning in the space surrounding the comet, like a swarm of gnats sizing up a grapefruit. A game of baseball would take years to complete – which is just as well, since you have a million years or so to idle away. But the ground rules would be unorthodox.

You pack the snow into an odd dark snowball and easily fling it off the comet, never to return again. With a flick of the wrist, without even engaging your arm in a throwing motion, you have launched a new comet on its own long, falling trajectory into the inner solar system. At the equator of your comet, you can lightly pitch a snowball in the direction opposite to the spin of the comet, so that its rotation is cancelled and it hovers forever above the same point on the surface. You can make arrays of objects stationary in space – vast, apparently motionless three-dimensional assemblages, poised above the cometary surface.

As the millennia pass, you cannot help but notice that the yellow star is gradually growing more intense, until it has become by far the brightest star in the sky. The early phase of your voyage has been tedious, even if you are endowed with heroic patience, and in several million years hardly anything has happened. But you can at least see your surroundings more clearly now. The icy ground beneath you has barely changed. The journey has been so long that you have been able to detect variations in the brightness and, you think, even the positions of several nearby stars. Your world is moving faster now, but otherwise everything is still, silent, cold, dark, changeless.

The comet eventually begins crossing the orbits of other kinds of objects, much larger bodies that are also bound in gravitational thrall to that beckoning point of light. Below, great worlds of gas swing ponderously into view. As you pass close by them, you career perceptibly. Their gravities are so large that they retain massive atmospheres. Your comet, by contrast, is so insubstantial that any puff of gas released escapes almost instantly to space. Accompanying the giant gas planets with their multicoloured clouds are retinues of smaller, airless worlds, some of them made of ice – much more akin to the comets than the huge ball of hydrogen that fills your sky.

You can feel the warmth of the Sun increasing. The comet feels it too. Little patches of snow are becoming agitated, frothy, unstable. Grains of dust are being levitated over the patches. Considering the feeble gravity, it is no surprise that even gentle puffs of gas send grains of ice and dust swirling skyward. A powerful jet gushes up from the ground and a fountain of fine particles is launched far above you. The ice crystals sparkle prettily in the sunlight. After a while, the ground becomes covered with a light snow. As you plunge onward, closer to the Sun, its disc now easily visible, such blowoffs become more frequent. While on one of your excursions aloft, you chance to see an active jet, a geyser pouring out of the ground. You give it a wide berth. But it reminds you of the instability – the literal volatility – of this tiny world.

Far out in space, the vanguards and outriders of the columns of crystals are being blown back by some invisible influence. Eventually, the cometary nucleus on which you have been riding is enveloped in a cloud of dust particles, ice crystals and gas, and the material being blown back behind you slowly forms an immense but graceful tail. If you stand on solid ground, far from the kinds of unstable ices that produce the geysers, you can still see a fairly clear sky, and track your motion by the stars. When the big jets go off, you can feel the ground shift. Here and there, the ice has sheared or cracked or fallen, revealing intricately stratified layers of various colours and darkness – a historical record of the building of the comet from interstellar debris billions of years ago. By looking at the stars, you can tell that your worldlet is darting erratically, rebounding in the opposite direction every time a new gusher erupts. The fountains of fine particles cast diffuse shadows on the ground, and there is now a sufficient number of them – most still of modest dimensions – that the field of darkish ice

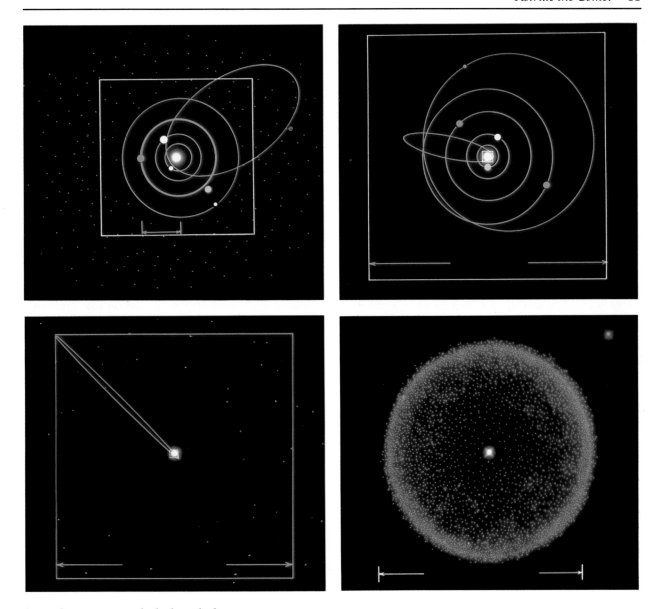

has taken on a mottled, dappled appearance.

The fine, icy grains evaporate in a few moments when heated by the increasingly fearsome sunlight, and only the refractory, involatile dark grains that they contain will persist as solids. The substance of the comet is being converted into gas before your eyes. And the gas, illuminated by the sunlight, is glowing eerily. You realize that there is now not just one tail, but several, each produced by its own geyser. There are straight blue tails of gas, and curving yellow tails of dust. No matter in what direction the jet happens to gush at the beginning, the unseen hand carries it away from the Sun. As the jets turn on and off, and the streamers above you curve because of the rotation of the comet, a rococo skywriting takes form. But everything aloft is relentlessly redirected by invisible forces, the pressure of light and the wind from the Sun. The solar wind seems intermittent. So the gradually burgeoning tails form, merge, separate, and dissipate, and knots of higher brightness abruptly accelerate and then decelerate leeward from the Sun. And windward, sheets of gas and fine particles form complex and exquisite veils that change their aspect in the twinkling of an eye. This is a kind of polar fairyland, and its beauty momentarily distracts you from how dangerous it has become.

Because of the evaporation of so much ice, the ground near the exhausted gushers is friable, delicate, fragile, often no more than a matrix of fine particles stuck together billions of years ago. Soon, hills of snow that had stolidly resisted the importuning sunlight show signs of stirring. There is an internal motion. The ground buckles. Tentative puffs of gas are released to space, then many geysers simultaneously spew, and you know that nowhere on the entire surface of this cometary nucleus is there a safe refuge. The comet has awakened from a four-billion-year trance into a wild and manic frenzy.

After you pass the Sun and retreat into the interstellar night, the comet will lose its tail and settle down. Its orbit will one day carry it back again to the inner solar system. Perhaps in some future pass by about the Sun, millions of years hence, the surface will be safer because all the outer layers of ice will have already been vaporized by the heat. After many passages, a comet will become less active, produce fewer jets, generate a less spectacular tail. As they get older, they settle down. But you are aboard a new comet, and swirling geysers are spraying the skies with dust. Maiden voyages are always the most dangerous.

You are venturing still closer to the Sun and, although the sky above you is overcast, the heat becomes stifling. The surrounding haze of fine, bright particles that diffuses the sunlight for eons, also reflects it back to space, and if not for this protection the cometary nucleus and you yourself might become uncomfortably warm – even, red hot.

You are rounding the Sun now, racing, hurtling through this treacherous regime. You have never moved this fast before. The ground is creaking and straining, new fountains are violently erupting. You take refuge at the shadowed base of a hillock of ice whose sunward side is crumbling, evaporating, and shooting pieces of itself out into space. But eventually the activity subsides, and the skies partially clear. Formerly, the Sun was ahead of you; now it is behind, on the other side of the nucleus.

Through a break in the surrounding nebulosity you realise that you are passing close to a small, blue world with white clouds and a single battered moon. It is the Earth in an unknown epoch. There may be beings there, perhaps, who will look up and see this apparition in their skies, who will note the great blue and yellow tails streaming away from the Sun, the complex pinwheel-shaped patterns of fine-grained material being jetted off into space, and they will wonder what it means. Some of them might even wonder what it is.

It is surprising how closely you are passing, and it occurs to you that sooner or later some comet is going to run smack into this little planet. The Earth would survive such an impact, of course – although there would doubtless be minor attendant changes, some species of life coming to an end, perhaps, and others beginning anew. But the comet would not survive. It would fall deeper and deeper into that atmosphere, large fragments, whole hills of the cometary nucleus separating off, flames licking through the crevices into its hidden interior. Perhaps enough of the comet would survive to make a huge explosion, generating a large hole in the ground down there, and spraying up a cloud of surface dust. But of the comet itself, all the ices would have vaporized. All that would be left would be a sprinkling of fine, dark grains, scattered like birdseed or buckshot upon this alien land.

But, you reassure yourself, running into a world is an unlikely event. On this swing past the Sun, at any rate, you will not collide with anything larger than occasional motes of interplanetary dust, the remnants of past comets that have spent their substance dashing through the realm of fire. With a last glance at the blue world, you silently wish its inhabitants well. Perhaps they will consider their skies comparatively drab and cheerless once your comet has gone. For yourself, you are happy to be on the return trajectory, out of the deadly heat and light, and back towards the placid cold and dark, where, except for an occasional unlucky jostling by a passing star, comets can live forever. The long and lovely tails are now preceding you on your journey; the wind from the Sun is at your back.

Chapter 2

PORTENT

If at your coming princes disappear,
Comets! come every day – and stay a year.

Samuel Johnson,
Letter to Mrs Thrale
(6 October 1783).

For a million years or more, there have been human observers to wonder at the grandeur of the comets that intermittently grace the skies of Earth. Memorable comets, visible to the naked eye, have appeared on average once a decade, a few times a lifetime. On a hundred thousand occasions during the tenure of humans on Earth, our ancestors must have seen a diffuse streak of light, brightening night after night, sometimes outshining even the brightest stars, and then, weeks or months later, slowly fading from view. A hundred thousand apparitions – and yet we have not a single concrete recollection of any comet earlier than the last three or four millennia. In this sense, as in many others, we are a species of amnesiacs. We are estranged from our past. We can only guess what our ancestors thought when the habitual serenity of the heavens was so spectacularly interrupted.

For most of that time, humans enjoyed a far more intimate relationship with the sky than we do now. We may know much more than they did about what is actually up there, but they were much more a part of it. They slept among the stars. They looked to the sky to tell them when to camp and when to move on, when to expect the migratory game and the rains and the bitter cold. Appropriately, they watched the sky as if their lives depended on it, but also because it was a puzzle whose complex beauty moved them. They invested the heavens with hypotheses, explanations, and metaphors that we call myth. The apparition of a bright comet is a recurring occasion for marvel and speculation, and thus the comets prodded us along the road to consciousness.

Comets must once have been a kind of psychological projective test – something wholly unfamiliar that you must describe in ordinary language. The Tshi people of Zaire call comets 'hair stars', and the word comet – the same in many modern languages – comes from the Greek word for hair. A comet suggests flowing tresses. To the Chinese, comets were 'broom stars', and much else. In other cultures, they are 'tail

2.1 *(Opposite) Twelve representations of comets from 168 BC to the present, including depictions from China, Turkey, Germany, Italy, Japan, France, pre-Columbian Mexico and an American spacecraft. (Anne Norcia.)*

2.2 *The tails of many comets suggest flowing hair. This is a drawing of Donati's Comet on 29 September 1858, as seen through the telescope by G.P. Bond, Harvard College Observatory (From Amédée Guillemin, The Heavens, Paris, 1868.)*

stars', or 'stars with long feathers'. Scientists today still describe comets as having 'tails'. To the Tonga, comets are 'stars of dust', which is much closer to the truth. The Aztecs saw them as 'smoking stars'. Among the Bantu-Kavirondo, all apparitions are returns of the same comet: 'There is but one comet, Awori, the feared one, with his pipe.'

It was hard not to take the comets personally. Nearly everybody did. This tendency to construe each apparition as a telegram to the locals from the gods permeates virtually every cometary record until the sixteenth century. Rarely have so many diverse cultures, all over the world, agreed so well. In the history of the world, more societies have advocated incest or infanticide than have taught that comets were benign, or even neutral. Everywhere on Earth, with only a few exceptions, comets were harbingers of unwanted change, ill-fortune and evil.

In their myths, the tribal peoples of Africa may have preserved something of our original perceptions of comets. To the Masai of East Africa, a comet meant famine; to the Zulu of the South, war; to the Eghap of Nigeria, pestilence; to the Djaga of Zaire, specifically smallpox; and to their neighbours, the Luba, the death of a leader. The !Kung of the Upper Omuramba in what is now Namibia were alone in their optimism. They saw the comet as a guarantee of good times ahead. This is such an unusually cheerful interpretation that you might well ask who the !Kung are. (The '!' stands for a click you make by touching your tongue to the roof of your mouth at the same instant you say the 'K'. It takes some practice.) They are hunter-gatherers, with a rich culture that is closer to the long-term human norm than almost any other culture today. So you wonder whether the !Kung knew something about nature – or themselves – that the rest of us have forgotten.

There is an overwhelming sadness to the literature of comets. With melancholy consistency, we discover that disaster has always been commonplace; that any comet, at any time, viewed from anywhere on Earth, is assured of some tragedy for which it can be held accountable. This connection of comets and misfortune is made in the earliest surviving reference to a comet*, a single Chinese sentence from the fifteenth century BC:

> When Chieh executed his faithful counsellors, a comet appeared.

*And the most recent. Thirty-five hundred years later, comets continue to be linked with catastrophe. In Chapters 15 and 16, we discuss the current scientific debate over whether comets are responsible for the extinction of the dinosaurs and most other species. Even 'apparition' – the word we use today to describe the presence of a comet in our skies – is an echo, with ominous and supernatural overtones, of our ancient belief about comets.

These words, associating official perfidy and murder with the early stirrings of astronomy, were written two centuries before the birth of Moses. Three hundred years later, another writer noted:

> When King Wu-wang waged a punitive war against King Chou, a comet appeared with its tail pointing towards the people of Yin.

The Yinnites were in trouble.

SMALL ARBITER OF HUMAN DESTINY

Autumn orchids, luxurious jungle
Spread life below the hall.
Green leaves, white blossoms
A rich and fragrant scent overtakes you.
From every person comes lovely children.
Why, then, my Lord, such bitter sorrow?

Autumn orchids, fresh and lush
Green leaves, purple stem.
The hall is filled with lovely people.
Suddenly alone with me, a meaningful glance.

He came without a word, went without saying goodbye
Riding the whirlwind, carrying the cloud banner.
Grief beyond grief is life's separations.
Joy beyond joy are new friends.

Lotus garments, sweet basil belt
Suddenly came, hurriedly left.
His evening lodgings, the gods' frontier.
You wait for whom at the clouds' edge?

With your lady, bathe in the Pool of Union.
Dry her hair under the Sun.
I search the sky for my lovely one – why is he not yet here?
Face to wind, indistinct, I lift up my voice in song.

Peacock-feather canopy, kingfisher-blue flag
Climb up to the Ninth Heaven, soothe the comet.
Grasping his long sword, he projects and nurtures the young.
My Lord alone is fit to bring justice to mankind.

Chu Yuan (340-278 BC)
Translated by Heather Smith
and Xie Yong

2.3 Chinese comet calligraphy. Translated, it reads 'broom star'.

Chu Yuan, statesman and one of China's most beloved ancient poets, committed suicide while in exile by drowning himself in a river. Every 5 May, people in China still throw specially prepared rice into the rivers in a symbolic placatory gesture to prevent the fish from devouring his body. In this poem, the comet is both a metaphor for the lost lover and a link to a wise and compassionate god. Many of the seemingly irrelevant images – such as 'cloud banner' and 'blue flag' – are in reality allusions to the numerous Chinese names for comets.

2.4 *A portion of the world's first cometary atlas, the Mawangdui silk, from circa 300 BC*

2.5 *A Japanese Tsuba, or swordguard, from the seventeeth century, illustrating a comet at the Sumiyoshi Shrine. The comet can be seen just below the crescent moon at the top. (Courtesy, David Pepper/Okamé, Toronto.)*

If there was unanimity that comets caused disasters, there was disagreement as to what to do once you saw one. In Master Tsochhiu's *Enlargement of the Spring and Autumn Annals*, written between 400 and 250 BC, there is an entry entitled, 'Yen Tsu Argues Against the Use of Prayer in Averting Disaster from Comets':

> In the year [516 BC], a comet appeared in the kingdom of Ch'i. The king of Ch'i wished to send his ministers to pray to heaven. Yen Tsu dissuaded the king, saying, 'It's useless. You're only fooling yourself. Whether heaven will give you a disaster or good fortune is set; it won't change. How can you expect prayer to change anything? A comet is like a broom: it signals the sweeping away of evil.* If you have not done anything evil, why do you need to pray? If you have done something evil, praying won't avert disaster. The work of the Ministers of Prayer won't change fate.'
>
> The king was delighted to hear this, and ordered the praying stopped.†

A survey of the early history of cometary observation gives the impression that for almost a thousand years everybody, except the Chinese, went to bed early. The Chinese logged at least 338 separate apparitions from roughly 1400 BC to 1600 AD. Since 240 BC they have missed a return of Halley's Comet only once, in 164 BC. Their neighbours, the Koreans and the Japanese, made valuable but far less frequent observations. In the West, nothing approaching systematic observation of comets was in evidence until the fifteenth century.

In the 1970s, an excavation of the 'Number Three Tomb' at Mawangdui, near Changsha, revealed the most impressive example of Chinese priority in this field – an illustrated textbook of cometary forms, painted on silk. Part of a larger work concerning clouds, mirages, halos and rainbows, it was compiled around 300 BC. Twenty-nine comets are displayed, classified by their appearance and by the particular brand of mayhem each foretells. Eighteen of the thirty-five different names known in Chinese for comets are given here. The four-tailed comet signifies 'disease in the world', the three-tailed one, 'calamity in the state'. A comet with two tails that curve to the right promises a 'small war', although, at least, 'the corn will be plentiful'. (This notion that the sole salutary effect of comets is as a boon to agriculture was also shared by medieval Anglo-Saxons.)

How long would it take to assemble a catalogue of twenty-nine distinct cometary forms? With 338 separate sightings recorded over 3,000 years in the surviving Chinese annals, the average discovery rate is about one bright comet, visible to the naked eye, a decade – not too far from current

---

*This is a pun. One of the many Chinese ideograms for comet is 'broom star'.
†Translated by Heather Smith and Xie Yong.

RECORD OF THE WORLD'S CHANGE

Comets are vile stars. Every time they appear in the south, something happens to wipe out the old and establish the new. Also, when comets appear, whales die. In Sung, Ch'i and later Ch'in times, when a comet appeared in the Constellation of the Big Dipper, all soldiers died in chaos...

When a comet appears in the North Star, the Emperor is replaced. If it appears in the end of the Big Dipper, everywhere there are uprisings and war continues for several years. If it appears in the bowl of the Dipper, a Prince controls the Emperor. Gold and gems become worthless. Another explanation: Scoundrels harm nobles. Some leaders appear, causing disturbances. Ministers conspire to rebel against the Emperor...

When a comet travels north but points south the country has a major calamity. Western neighbours invade and later there are floods. When a comet travels east and points west, there are uprisings in the east.

...When a comet appears in the Constellation Virgo, some places are flooded and there is severe famine. People eat each other... If the comet appears in the Constellation Scorpio, there are uprisings, and the Emperor in his palace has many worries. The price of rice goes up. People migrate. There is a plague of locusts.

...When a comet appears in the Constellation Andromeda, there are floods and migrations of people. Many rise up and the country is divided by civil war. When a comet appears in the Constellation Pisces there is first drought and later flooding. Rice is expensive. Domesticated animals die and an epidemic strikes the army.

When a comet travels into the Constellation Taurus, in the middle of the double month*, blood is shed... dead bodies lie on the ground. Within three years the Emperor dies and the country is in chaos. When a comet appears in Orion there are major uprisings. Princes and Ministers conspire to become Emperor. The Emperor has many worries. Everywhere there is disaster by war...

When a comet appears in the Constellation Hydra, there is war and some conspire to overthrow the Emperor. Fish and salt are expensive. The Emperor dies. Rice also becomes expensive. There is no Emperor in the country. The people hate life and don't even want to speak of it.

Record of the World's Change.
Li Ch'un Fung, 602-667 AD
Translated by Heather Smith and Xie Yong.

*The Chinese calendar at this time was a lunar one, but in every year there was a 'double month' to adjust to solar time.

2.6 *As late as 1528, European perceptions of comets had distinct hallucinatory elements. In this woodcut of the comet of that year, the comet is seen as a melange of decapitated heads and miscellaneous implements of warfare. The picture was based on the description of this comet by Ambrose Paré. (Courtesy, J. Veverka.)*

values. If each of the twenty-nine forms appears equally often, you would have to wait 290 years to see them all. But some cometary forms are much rarer than others. Thus, if every form depicted corresponds to a different comet, the Mawangdui Atlas must have drawn on an earlier continuous tradition of systematic observation preceding it by many centuries, possibly millennia. Accordingly, this splendid tradition of recording cometary forms must date back to 1500 BC or earlier. The earliest written and the earliest graphic representations of comets trace back to, or at least through, the same epoch. Perhaps there was an out-of-the-ordinary comet then that commanded their attention.

Our admiration for the Mawangdui silk is further enhanced when we consider the depictions themselves, which are at least roughly consistent, in a few bold strokes, with modern photographs of the comets. These observers drew what they saw. We have only to compare these images with a European woodcut of the comet of 1528 (Figure 2.6) to appreciate their sobriety: no dragons; no swords; no devils. Just comets.

In surveying the longstanding Chinese absorption with comets, we are reminded of another area in which the Chinese led the way – the invention of fireworks. Might the ancient Chinese, we wonder, have devised rockets to adorn the heavens during the long and tedious intervals between comets? Even if rockets and comets were not connected then, they certainly are today (see Chapters 6 and 18).

The ancient Chinese assembled a large, accurate, and detailed body of data on comets. Their catalogues list, for hundreds of apparitions, many of the following pieces of information – a date, the kind of comet, the constellation in which it was first seen, its subsequent motion, its colour and apparent length, and how long before it disappeared. Sometimes, day-by-day changes in the length of the comet are recorded. But, for all this, they never had an inkling of what comets really were. This was to be wholly an achievement of the West, although it was a long time in coming. Western cometary astronomy prior to the Renaissance is a chronicle of occasional episodes of lucidity – especially in Ionia, Athens and Rome – punctuating a far longer and more widespread gloom of ignorance, superstition and delusion.

The earliest unambiguous references to comets in the West come from what is today Iraq. The few surviving Babylonian fragments remind us of their African and Chinese counterparts. Consider this, from the time of Nebuchadnez-zar I, in the twelfth century BC:

> When a comet reaches the path of the Sun, Gan-ba will be diminished; an uproar will happen twice...

The diminishment of Gan-ba is bad news, you can be sure. Occasionally, as in this notice from the same time and place, the auspices are favourable:

> When a star shines and its brilliance is as bright as the light of day, [when] in its shining it takes a tail like a scorpion, it is a fortunate omen, not for the master of the house, but for the whole land.

A comet that rivalled the Sun in brightness must have come very close to the Earth.

The confidence with which these ancient astrological pronouncements are made is striking. There is no hedging of bets, no ambiguity, and no curiosity. We never find a confrontation of two contending hypotheses, much less an appeal to observations to decide the issue. There is always a Gan-ba in the picture. Science had not yet been invented.

In the works of Diodorus of Sicily (*c* 60-21 BC) and Lucius Annaeus Seneca of Rome (*c* 4 BC-65 AD) there is indirect evidence – or perhaps it is only hearsay – that the Egyptians and Babylonians had devised some scientific understanding of the comets. Diodorus wrote:

> As a result of their long observations, they [the Egyptians] have prior knowledge of earthquakes and floods, of the rising of comets, and of all things which the ordinary man looks upon as beyond all finding out.

The annual times of flooding of the Nile Valley were well-known to the ancient Egyptians. From the odd behaviour of animals, it is possible – as the modern Chinese have demonstrated – to predict an earthquake early enough to save many lives. But correctly foretelling the apparition of a comet is much more difficult. Maybe someone made a lucky guess.

Seneca reports an opinion, which he does not share, that the Babylonians believed comets were bodies something like planets. There is no elaboration. We do know that the Egyptians and Babylonians made seminal contributions to mathematics. However, it was in Greece in the fifth century BC that curiosity turned away from the supernatural, and first found its world-changing means of expression: science.

Everything we know of the inventors of this new way of thinking comes to us second-hand. Democritus (born around 460 BC and believed to have lived to a very old age) wrote at least seventy works, all of which have been destroyed or lost. Our knowledge of Democritus comes mostly from Aristotle (384-322 BC), who held him in high regard and disagreed with virtually everything he had to say. We are told that Democritus believed comets to be produced when one planet passed near to another. Democritus may have properly distinguished star from planet, but even if not, he was on the right track: comets, he was saying, are celestial bodies and arise by natural processes. So far as we know, no one had ever proposed this outlandish idea before.

Aristotle believed he could disprove it by noting that Jupiter had, in his time, come close to a star in the constellation Gemini, and had not produced a comet. But Aristotle did not know that the star was light years behind the planet, and that only from our perspective were they passing 'near' one another. In this matter, as in most others, Democritus was the

But after the conquest of Troy and the annihilation of its descendants… overwhelmed by pain she separated from her sisters and settled in the circle named arctic, and over long periods she would be seen lamenting, her hair streaming. That brought her the name of comet.

Hyginus,
De Astronomia (*c* 35 BC,
translated from the
Latin by Milad Doueihi).

better astronomer. But Aristotle's argument appealed to observation, not myth or conventional wisdom. The debate was scientific. This also seems to be the earliest mention of the possibility that comets are disgorged by Jupiter; the idea has had nine lives at least, and was, not so long ago, a cause of vigorous controversy in the United States.

Aristotle's reasons for believing that comets could not live among the planets was based partly on observation. He framed a supportable scientific hypothesis, which went something like this: the zodiac is the succession of constellations, many named after animals, through which, over the months and years, the planets, the Sun and Moon move. (During daylight, you can't see what constellation the Sun is in, of course, but with a star map at twilight or dawn you can.) The zodiac runs all the way around the sky at an angle to the horizon. For all our ancestors knew, the planets and the Sun and Moon might, over the course of a lifetime, wander through every constellation in the sky. Since this does not happen, all the planets must lie very nearly in the same plane. In contrast, comets are observed to travel sometimes within, but sometimes well outside the zodiac. Furthermore, unlike planets, comets change their forms in a few days, before the eyes of the observers. Thus, comets could have nothing in common with planets. They must be sublunary, beneath the Moon – that is, within the Earth's atmosphere. (Aristotle thought the Moon represented the farthest reach of the atmosphere.) The conclusion was clear: comets were a form of weather. Although there was a little debate early on, this view held sway for 2,000 years.

All of Aristotle's astronomy was predicated on his deep conviction that the heavens were 'free from disturbance, change and external influence...' He believed that the Earth was absolutely stationary in space – as if nailed down. The heavens, on the other hand, were whipping around the Earth at the brisk rate of one rotation per day. The bottom of the atmosphere clearly is stationary with the Earth. But the top of the atmosphere must share the sky's rotation. Now, imagine an exhalation of hot, dry gas from the Earth – perhaps through a fissure, a crevice, or a volcano. That gas will rise and, when it reaches the sky, be heated by the Sun until it bursts into flame. But since the burning gas has reached the realm of the heavens, it must now move with the stars and the planets. This was Aristotle's explanation of the comets. And, given the limitations of the science of his day, it was far from foolish.

He thought the aurora borealis, and even shooting stars, were examples of the same sort of phenomena – exhalations from the interior of the Earth rising to the stars. Comets survive, he taught, until all the gas had burst into flame. New comets were created by new exhalations. There was thus a balance or steady state between the production and the destruction of visible comets, an idea still central to understanding them. Aristotle held that there are so few comets because most of the flammable vapours exhaled by the Earth were otherwise employed – in producing the continuous band of fire called the Milky Way. In contrast, Democritus had concluded that the Milky Way is composed of an enormous number of stars, so far away that we cannot see them individually – exactly the right answer.

Aristotle was forced to concoct a terrestrial source for the comets, since he had locked them out of his changeless skies, in which, for quasi-religious reasons, he had decreed that no new celestial bodies may be born and no old ones die. His insistence on the immutability of the heavens was the most influential error in the history of astronomy, contributing to a detour from reality that lasted nearly two millennia. But Aristotle cannot be held fully accountable for the credulous acceptance accorded his opinions by succeeding generations.

Seneca, born in Cordoba, Spain, to a wealthy and famous family, was a contemporary of Jesus, although the two never met. His brother, though, was an acquaintance of St Paul. As a young man, Seneca came to Rome, where he studied grammar, rhetoric, law and philosophy. He enjoyed a substantial reputation as writer and orator until the year 41, when he was banished to Corsica for sleeping with Caligula's sister. Considering this emperor's predilection for cruelty, the punishment was mild. Seneca spent his years of exile writing, and studying philosophy and natural science; astronomy was among his pursuits.

In the year 49, he was recalled to Rome to teach. As a tutor his success was mixed, his only pupil being the future emperor, Nero. When, at the age of seventeen, Nero was raised to the purple, Seneca became political advisor to the Emperor and minister of state. For the next eight years, Seneca and Sextus Afranius Burrus, the commander of the Praetorian Guard, ran the Roman Empire. By all accounts they performed well, fostering fiscal and judicial reforms, and somewhat easing the lot of the slaves. But Nero grew more tyrannical, Burrus died – perhaps as a result of foul play – and Seneca's political power withered. He withdrew from public life, and wrote some of his most celebrated works until, in the year 65, he received an imperial command to commit suicide for his alleged involvement in a conspiracy against the Empire. He died with courage and composure.

Seneca left works on many subjects, but it is his *Natural Questions*, written during the last years of his life, that concerns us here. The seventh 'book' is entitled 'Comets', and

2.7 *Bust of Lucius Annaeus Seneca of Rome (c4 BC-65 AD).*

Seneca does the subject considerable justice, taking on Aristotle with some success. He argues that comets could not be atmospheric disturbances: they move with stately regularity, and are not dissipated by the wind. Thus,

> I do not think that a comet is just a sudden fire, but that it is among the eternal works of nature.

In refuting Aristotle's argument that comets cannot be planets, since they are not restricted to the zodiac, Seneca asks, faintly echoing Job:

> Who places one boundary for planets? Who confines divine things in a narrow space? [The planets]... have orbits that are different from one another. Why, then, should there not be other stars which have entered on their own route far removed from them? What reason is there that in some part or other of the sky there should not be a passageway?

This one description ought to be generally agreed upon: [a comet is] an unusual star of strange appearance... seen trailing fire which streams around it.

Seneca,
Natural Questions
Book 7, 'Comets'.

To the objection that stars can be seen through comets, so comets must be incorporeal and cloud-like, Seneca replies correctly that the transparency applies only to the tail, not necessarily to the head.

One of the most fascinating passages is an exposition and critique of the views of one Apollonius of Myndos, an otherwise unknown Greek scholar from the fourth century BC:

> ...Many comets are planets... a celestial body on its own, like the Sun and the Moon. It has a distinct shape... not limited to a disc, but extended and elongated lengthwise... A comet cuts through the upper regions of the universe and then finally becomes visible when it reaches the lowest point of its orbit... Comets are many and various, different in size, unlike in colour... Some are bloody, menacing – they carry before them the omen of bloodshed to come. Others diminish and increase their light, just as the other stars [planets] do which are brighter and larger when they descend because they are seen from a closer position, and are smaller and dimmer when they recede because they are withdrawing far away.

Except for the business of comets as omens, Apollonius' views seem astonishingly modern. But Seneca will buy none of it.

In 'Comets', Seneca's style is so direct and contemporary that you have no difficulty hearing his voice speaking inside your head. 'We can only investigate these things,' he tells you, 'and grope in the dark with hypotheses, not with the assurance of discovering the truth, and yet not without hope.' He continues:

> Many things that are unknown to us the people of a coming age will know. Many discoveries are reserved for ages still to come... Nature does not reveal her mysteries once and for all. We believe we are her initiates, but we are only hanging around the forecourt...

And the loiterers in the temple forecourt were dwindling. Seneca laments what he saw in his time as a waning of 'interest in philosophy. Accordingly, so little is found out from these subjects which the ancients left partially investigated, that many things which were discovered are being forgotten.' For good reason, he surmised a growing intellectual lassitude in his world. His reliance on reason, his willingness to decide between alternatives on the basis of objective evidence, stand in marked contrast to the approach to comets of the next generation – of Lucan (39-65), for example, Seneca's own nephew:

> The heavens appeared on fire, flaming torches traversed in all directions the depths of space; a comet, that fearful star which overthrows the powers of the Earth, showed its horrid hair.

Or consider the view of the historian Pliny the Elder (23/24-79):

> A fearful star is the comet, and not easily appeased, as appeared in the late civil troubles when Octavius was Consul; a second time by the... war of Pompey and Caesar; and, in our own time, when, Claudius Caesar having been poisoned, the empire was left to Domitian, in whose reign there appeared a blazing comet.

Lucan and Pliny were the wave of the future.

Josephus mentions in his *History of the Jews* that a 'sword' hung over Jerusalem for the whole year, foretelling the destruction of the city in the reign of the Emperor Vespasian. This is probably a reference to the apparition in the year 66 of Halley's Comet. But how could a natural object hang over a city for a year? The Earth turns. A comet would rise and set with the stars, and surely cannot be described as 'hanging' over any place on the Earth's surface. Meteors streak and vanish in an instant, planets cannot look like a sword, the aurora borealis is too far north, and even an artificial satellite cannot hover over non-equatorial Jerusalem. Barring a miracle, Josephus' sword – unrecorded by any other chronicler of the time – must, along with other matters in his chronicle, be considered with scepticism.

As classical learning dwindled over the centuries, the triumph of superstition was symbolized by the deaths of two emperors. The comet of the year 79 was said to portend the death of the Emperor Vespasian, who was already involved with cometary portent, and a shameless employer of divine omens to justify and authenticate his seizure of the Empire. Vespasian, however, expressed a healthy scepticism. 'That hairy star does not portend evil to me,' he said. 'It menaces, rather, the King of the Parthians,' his long-time adversary. 'He is a hairy man,' Vespasian concluded, 'I am bald.' His

2.8 *A seventeenth century woodcut of a comet hanging over Jerusalem, according to the account of Josephus. From* Theatrum Cometicum *by Stanislau Luvienietski, Amsterdam, 1668. (From the collection of D.K. Yeomans.)*

2.9 *A representation of a comet as a sword; the figure was drawn in the manner of the description in Pliny, but at a later time.*

2.10 *A depiction of what would later be called Halley's Comet over the city of Nuremberg in 684. The woodcut was done well after the event. (From* The Nuremberg Chronicle. *Courtesy, Library of Congress.)*

scepticism did not save him, though, and he died the same year. By the early Middle Ages, the association of comets with the deaths of princes was so deeply ingrained that when a ruler died and no comet was seen – as was the case at the death of Charlemagne in 814 – a great comet, otherwise invisible, was generally acknowledged to be passing by the Earth. The Emperor's new comet.

You may steep yourself in the cometary discourse of the whole of the Middle Ages, and still find no one within shouting distance of Democritus, Apollonius, Aristotle, or Seneca. The medieval treatises are full of divination and portent, omen and blood, mysticism and superstition, with hardly ever, even as a passing thought, the view that comets might be just a part of nature, and not a warning to the wayward. Even the historian Isidor, Bishop of Seville (602-636), who denounced astrology and astrologers, believed that comets presaged 'revelations, wars, and pestilence'. For more than a millennium, there was no dissenting opinion, and even the occasional attempt at purely factual statements – such as the announcement by the Venerable Bede (673-735) that comets are never seen in the western sky – were often wholly in error on straightforward and often verifiable matters of observation.

So, when the Renaissance and the Enlightenment finally came, a new breed of scholars arose who were predisposed to hold the Church responsible for superstition and ignorance – about comets, and many other matters. Callixtus III was a Spanish-born Borgia, untainted by scandal, who was made Pope in 1455 at the age of 77. It was the obsession of the end of his life to recover Constantinople from the Turks, the adversary ideologies of the time being as fervent and narrow as the deadlier struggle in our own day between capitalism and communism. Each side confidently called upon the One True God, and publicly proclaimed the inevitability of victory.

In 1456, a great comet appeared in European as in Chinese skies. Once detected, the comet seems to have struck real terror in beleaguered Christendom. We know today, as Callixtus and his contemporaries did not, that what they were seeing was the periodic return of Halley's Comet. Callixtus was convinced, so the story goes, that the visitor was an evil omen somehow allied with the Turkish cause; accordingly, he excommunicated it, ordering this heartfelt plea inserted into the *Ave Maria:* 'From the Devil, the Turk and the Comet, Good Lord, deliver us.'

According to many accounts, Callixtus sent 40,000 defenders to Belgrade, a city under Christian control besieged by the Turks, where, on 6 August 1456, with Halley's Comet hanging overhead, a great battle was fought which lasted two

2.11 *The comet of 1066 as depicted in the Bayeux Tapestry. King Harold of England, about to be overthrown by the Norman invaders, wonders whether to heed the portent.*

days. One later historian described the battle in these words:

> The Franciscans, unarmed, crucifix in hand, were in the front rank, invoking the papal exorcism against the comet, and turning upon the enemy that heavenly wrath of which none in those times dared doubt.

The armies of Mohammed II were repulsed, and both the comet and the Turks retreated. Constantinople was never recovered by the Christians.

One astronomical writer after another has quoted this account of the evenhanded anathematization of Turk and comet, and yet there is no record of such a prayer or curse in the Vatican archives, and no evidence that Callixtus excommunicated the comet, its previous adherence to the tenets of Christianity being at any rate in doubt. On 29 June, a Papal Bull was issued ordaining public prayers for the success of the Crusade; there is not a word in it about the comet, which was invisible to the naked eye by the middle of July, while the decisive victory over the Turks was a week or so later.

As far as can be determined, the story most astronomical writers relate about Callixtus and the comet traces back to *The System of the World* by Pierre Simon, the Marquis de Laplace, whom we will meet again in these pages. A brilliant scholar who left an enduring mark on the history of physics and astronomy, he was also a partisan of the French revolution, and its rationalist underpinnings. His *System of the World* was published 'in Year VII of the Republic' (1799). Because the Church had been intimately allied with the brutal regime of the Bourbon Kings, it is possible that Laplace was disinclined to be overly generous to Callixtus III. But Laplace did not make the story up, and the confusion seems to trace back to the work, *Lives of the Popes* (1475), by the Vatican Librarian, Platina. The image, however spurious, of a Pope solemnly excommunicating a comet was thought a plausible extension of the prevailing view of comets.

Far from the Vatican, and a generation later: in Tenochtitlan, the Aztec Emperor, Montezuma II (1466-1520), was awaiting the great white-bearded god, Quetzalcoatl, whom prophecy had decreed would return to Mexico and reclaim his Empire. When two brilliant comets arrived in rapid succession and appeared to rendezvous in the sky, Montezuma took them as a sure omen that Quetzalcoatl was on his way – that the Aztec Empire was his no longer. Withdrawn and disconsolate, he took every fire, storm, and oddity of nature as further warning. He was reduced to immobility by two comets and a prophecy. When in 1519, soon after the appearance of the comets, the white-bearded conquistador Hernan Cortes arrived out of the eastern sea with an expeditionary force of

2.12 *Montezuma I of Mexico is distressed to see an omen in his sky. (From* Los Tlacuilos de Fray Diego Duran.*)*

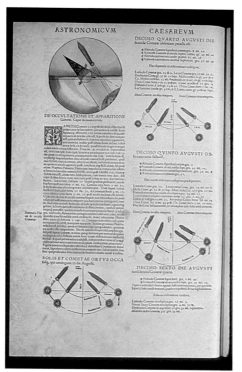

2.13 *The 1531 apparition of Halley's Comet, as depicted in the* Astronomicum Caesareum *of Apianus (1540). Note that the tail of the comet is correctly shown as pointing away from the Sun. (Courtesy, Library of Congress.)*

2.14 *A Dutch coin bearing the date 14 November 1577, and depicting the Great Comet of that year flying above the clouds and the landscape of the Earth. The Latin inscription reads: 'The Star of Offended Divinity'. (Courtesy, The American Numismatic Society, New York.)*

600 men and a few horses, Montezuma did not have to be much persuaded. He handed back the Empire to Quetzalcoatl. The Aztecs were helpless against Cortes' small band for more reasons than one. But the conquest and plunder of Mexico, and the annihilation of Aztec civilization, were in some significant measure due to a fatalistic dread of comets.

Across the Atlantic, the Protestant Reformation was soon underway. The contending sects, divided on many theological issues, were in perfect harmony on the matter of comets. On this issue at least, the movers and shakers of the Reformation saw eye-to-eye with Montezuma II. Martin Luther (1483-1546) spoke for all of them when, in an Advent sermon, he preached:

> The heathen writes that the Comet may arise from natural causes; but God creates not one that does not foretoken a sure calamity.

Andreas Celichius, the influential Lutheran bishop of Altmark, wrote in 1578 of comets as

> ...the thick smoke of human sins, rising every day, every hour, every moment, full of stench and horror before the face of God, and becoming gradually so thick as to form a comet, with curled and plaited tresses, which at last is kindled by the hot and fiery anger of the Supreme Heavenly Judge.

This view of comets as congealed sin ignited by the wrath of heaven, shows how many backward steps had been taken since the time of Aristotle, to whom Celichius is clearly indebted. A retort was provided by Andreas Dudith in the following year:

> If comets were caused by the sins of mortals, they would never be absent from the sky.

From all this, you would have never guessed that in the middle of the sixteenth century a revolution in human understanding of comets, and much else, was about to happen. In Denmark, in the very year that Luther died, just such a 'heathen' was born. His name was Tycho Brahe. With his artificial nose of gold or brass, his entourage of dwarfs, his legendary drinking parties, and his isolated island observatory, he was a somewhat atypical astronomer.

In Tycho's time, the unimpeachable authority on comets was still Aristotle. His doctrine that comets were confined to the Earth's atmosphere because the heavens were fixed and changeless was a cornerstone of the sixteenth century model of the universe, endorsed by secular and religious authorities alike. This was not a controversial issue. All knowledgeable experts agreed with Aristotle. The first real doubts were raised on a night in 1572, when Tycho looked up at the constellation

Cassiopeia and saw a star 'brighter than Venus', where no star had been before. The new star was far beyond the Earth's atmosphere, and is even now known as Tycho's Supernova – 'nova' is the Latin word for new. Evidently, and to nearly everyone's surprise, the heavens were not immutable. Aristotle and the Church had been wrong. The supernova of 1572 was a reveille for the astronomers of Europe and, soon, for the culture of the world.

Five years later, a great comet blazed across the skies of Europe to overturn decisively Aristotle's now tottering world view. Because the comet of 1577 was visible for an extended period, Tycho and his colleagues were able to share information and test each other's hypotheses. The supernova of 1572 had prompted Tycho to approach the comet as if it were an astronomical body, rather than an atmospheric disturbance.

If you look up at a comet, you see it against a backdrop of more distant stars. As time goes on, it may move from constellation to constellation, but in any given period of a few days, it seems frozen within its constellation, and rises and sets with the stars. Tycho asked himself how the comet would look if it were merely an atmospheric disturbance and close to the Earth, and also how it would look if it were a body like the planets or the stars, far from the Earth. He was able to hold two contradictory ideas in his head at the same time. Place your finger in front of your nose and alternately wink your left and right eyes. You will see the finger seem to move against the more distant landscape. Now move the finger to arm's length and wink again. The finger still moves, but less so. The apparent motion is called parallax: it is merely the change in perspective from left eye to right. The closer the finger is to you, the more parallax or apparent motion against the background; the further the finger is from you, the smaller is the parallax.

Tycho realized that this same principle can be applied to a comet – provided you could observe it from two widely separated observatories (see Figures 2.17 and 2.18). If the comet is very close to the Earth, the perspective will change greatly between the two observatories, and each observer will see the comet in front of a very different constellation. But if the comet is very far from the Earth, then both observatories will see the comet in front of the same constellation. Through parallax, it is possible to measure the distance of a comet from the Earth. Telescopes are not needed – only sighting markers and graduated circles with angular degrees marked off on them. Tycho lived in the last generation before the invention of the astronomical telescope, and his measurements could have been done – at least crudely – at any time in the previous few

2.15 *Tycho's Comet and Newton's Comet. The Great Comet of 1577 (left), as seen by Cornelius Gemma, and the Great Comet of 1680 (right), as seen by J.C. Strum. (From Amédée Guillemin,* The Heavens, *Paris, 1868.)*

2.16 *One of the first drawings of a comet as seen through the telescope. Shown are two different sightings of the Comet of 1618 on consecutive nights. The comet is moving through the constellations Orion (right) and the Big Dipper. (Drawn at the telescope by J.B. Cysat, from* The Mathematica Astronomica Ingolstaat, *1619. Courtesy, D.K. Yeomans.)*

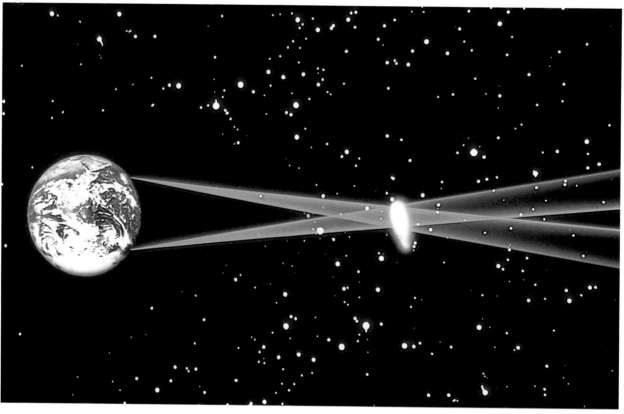

thousand years, had anyone thought to make the measurement.

Tycho was not the only one to make these measurements and do these calculations on the comet of 1577, and some of his contemporaries – perhaps still in the thrall of Aristotelian thinking – got the wrong answer, and deduced that the comet was in the Earth's atmosphere. Tycho's meticulous measurements and calculation, however, have stood the test of time. If the comet had been within the Earth's atmosphere, a sizeable parallax would have been detected. Tycho was able to find no significant parallax. With his precision of measurement, the comet of 1577 had to be much further from the Earth than is the Moon. The comet, therefore, must be somewhere up there among the planets and the stars. Employing international co-operation, elementary mathematics, and simple observations, Tycho found that conventional wisdom had been dead wrong for two millennia. If previous generations had known how distant the comets really were, perhaps they would have been less frightened by them. Tycho liberated the comets from the narrow circumterrestrial confines into which Aristotle had stuffed them, and allowed them to soar into space. Now, science was free to do the same.

The upheavals that mystics had always associated with comets was finally justified by the comet of 1577. In the vast literature of prophecy inspired by the comets up to this time, there is only one that commands our admiration. The prediction was made, not by a soothsayer or a priest, but by a scientist. It was Seneca:

> Why, then, are we surprised that comets, such a rare spectacle in the universe, are not yet grasped by fixed laws and that their beginning and end are not known, when their return is at vast intervals?... The time will come when diligent research over very long periods will bring to light things which now lie hidden... Some day there will be a man who will show in what regions comets have their orbit, why they travel so remote from other celestial bodies, how large they are and what sort they are.

Tycho set astronomy back on course; but the man thus prophesied was Edmond Halley.

(Opposite) If a comet were something in the Earth's atmosphere, as Aristotle maintained, then two observations from widely separated points would see it against quite different background stars. In the diagram above, 2.17, the blue cones represent the lines of sight of the two observers. But if the comet is very far away, 2.18 (below), the widely separated observers see the comet against almost the same stellar background. Simultaneous observations can therefore tell you something about the distance from the Earth to the comet. (Diagrams by Jon Lomberg/BPS.)

Who vagrant transitory comets sees,
    Wonders because they're rare; but a new star
Whose motion with the firmament agrees,
    Is miracle; for there no new things are.

John Donne,
'To The Countess of
Huntingdon', (1633).

Donne's astronomy was 56 years out of date; after Tycho, every comet as well as every supernova was a disproof of the Aristotelian view that in the heavens 'there no new things are'.

2.19 *Tycho Brahe's own drawing of the Comet of 1577 (at position X). In Tycho's scheme of the solar system, the comet went around the Sun, but the Sun and the other planets went around the Earth (A). (From Tycho Brahe's* De Mundi Aetherei, *1603. Courtesy, D.K. Yeomans.)*

EDMVND. HALLEIVS LL.D.
GEOM. PROF. SAVIL. & R. S. SECRET.

Chapter 3

HALLEY

[Mr Halley possessed] the qualifications necessary to obtain him the love of his equals. In the first place, he loved them; naturally of an ardent and glowing temper; he appeared animated in their presence with a generous warmth which the pleasure alone of seeing them seemed to inspire; he was open and punctual in his dealings; candid in his judgment; uniform and blameless in his manners, sweet and affable, always ready to communicate . . .

Jean Jacques d'Ortous de Mairan,
'Elegy for Mr Halley',
Memoires de l'Académie Royale des Sciences
(Paris, 1742).

When we think of Edmond Halley, if we think of him at all, it tends to be only in connection with his namesake – humanity's favourite comet. The comet becomes a kind of mnemonic device, operating at roughly seventy-five year intervals, that chides us into remembering him. To most of us, Halley is like the athlete who makes his way into the record books on the basis of one extraordinary season, or even one memorable play. We consult the records, expecting to find a journeyman, who, like countless others, has contributed a brick or two to the edifice of science. Instead, we find a master builder.

The date of his birth is uncertain; Halley believed it to be 29 October 1656. He began life in the Borough of Hackney, then a rural community outside London, but long since subsumed by the spread of the city. We have not a single anecdote, even apocryphal, that indicates the texture of his childhood. We do know that it was then that he first dreamt what he would later become. 'From my tenderest years I gave myself over to the consideration of Astronomy,' he recalled while still quite young. '[It gave me] so great pleasure as is impossible to explain to anyone who has not experienced it.' Halley's sense of science, not as livelihood but as rapture, was never to leave him. In his boyhood, two great comets were seen – one, in 1664, popularly associated with the Great Plague of London, the other in 1665, associated with the subsequent Great Fire. Although there is no record of Halley's having witnessed these comets, there is no way that a young person of his inclination and ability could have escaped being influenced by the comets of 1664 and 1665, so steeped by the popular culture of the time in portent of calamity and disaster.

3.1 *(Opposite) Edmond Halley, age thirty, by Thomas Murray. (Courtesy, The Royal Society of London.)*

3.2 *A contemporary horoscope of Edmond Halley. Halley had no use for astrology. The original horoscope is in the Bodleian Library, Oxford.*

His father, also named Edmond, was a businessman, a soap boiler and salter who owned lucrative properties in London. In 1666, the Great Fire devoured his real estate holdings, but his other businesses flourished. The recent horrors of the bubonic plague had given Londoners a new respect for personal hygiene. The manufacture of soap became a growth industry. Moreover, the expanding British Navy had a chronic need for salted meat to sustain its sailors on long voyages. The elder Halley, his business flourishing, was happy to use his recent wealth to see his son's evident promise realized. He sent him to St Paul's, one of the best schools in England, where Edmond did brilliantly. In 1671, he was elected Captain of St Paul's, which indicates that he was that rare thing – an excellent student who is popular with his classmates.

We know nothing of Halley's mother, other than that her name was Anne Robinson and that she died on 24 October 1672, nine months before Halley's departure for Queen's College, Oxford. Additional evidence of his father's largesse is provided by the quantity and quality of the astronomical instruments that Halley took along to college – among them, a twenty-four foot long telescope, which he immediately put to good use.

We know this because on 10 March 1675, the eighteen-year-old Edmond Halley had the audacity to write to John Flamsteed (1646-1719), England's first Astronomer Royal, informing him that the authoritative published tables on the positions of Jupiter and Saturn were in error. The young man had also found some errors in the star positions published by the incomparable Tycho Brahe. The tone of Halley's letter is not that of the young cowboy come to challenge the legendary gunslinger, but much more that of the youthful enthusiast, full of admiration for those who have come before him, anxious to join the club – and even more anxious to discover the true nature of the universe. We do not know exactly what Flamsteed's response was, but it must have been positive, because the next year he helped Halley publish his first scientific report, or 'paper'. It appeared in *Philosophical Transactions*, the journal of the Royal Society of London – then, as now, the leading scientific organization in Britain – and was titled *A Direct and Geometrical Method of finding the Aphelia, Eccentricities, and Proportions of the Primary Planets, without supposing equality in angular motion.*

What was this paper about? Since the work of Tycho's student, Johannes Kepler, it had been known that each planet moves along a path called an ellipse, a kind of stretched-out circle. The eccentricity is a measure of how stretched out the ellipse is: an ellipse with zero eccentricity is a circle, and an ellipse with an eccentricity of one or greater is not even a closed curve, but rather a parabola or hyperbola. The eccentricity of the Earth's orbit is 0.017, indistinguishable to the naked eye from a circle. Mercury's orbital eccentricity, by contrast, is 0.21 – perceptibly elongated. It would, many years later, be one of Halley's triumphs to determine that comets move in elliptical paths and to determine the eccentricity of cometary orbits, one of the keys to their origin.

In a roughly circular orbit, such as the Earth's, we are always pretty much at the same distance from the Sun. But in a highly elliptical orbit, like a comet's, the distance to the Sun varies depending on where the comet is in its orbit. The nearest point to the Sun, when the planet or comet is moving fastest, is called the perihelion (plural, perihelia). As the object sweeps around the Sun, it is said to be making 'perihelion passage'. The phrase has a nice ring to it, as if it were something experienced on a steamship of the old P&O line. The far point in the orbit is called aphelion. The more elliptical the orbit is, the bigger the difference between aphelion and perihelion. It is perfectly possible for a comet to have its perihelion near the Earth, and its aphelion far beyond the most distant known planet. In his first paper, Halley proposed a new and more accurate method of calculating the planetary orbits.

The paper had to be rewritten many times, in part due to Halley's inexperience, but also because the Bishop of Salisbury had published a contrary view, and so, Halley was told, might construe the paper as a personal insult. Halley had no intention of offending, and made the suggested changes happily. This was not the last time that Halley's pursuit of science would collide with ecclesiastical authority.

The same year that Halley's paper on orbits brought him to the attention of the world astronomical community, he chose to leave Oxford without taking his degree and to travel to the distant island of St Helena, west of Africa, to make the first map of the southern skies. The constellations you can see from high northern latitudes are almost completely different, of course, from those you see close to the South Pole. At the equator you see half of the northern and half of the southern constellations. St Helena was then the southernmost outpost of the British Empire and a good vantage point from which to observe not only the southern skies, but also some of the charted stars of the northern hemisphere – important, because if Halley's proposed map of the southern stars were to be useful to European astronomers, it would have to share some points of reference with star positions that had been previously determined. Beside its latitude, St Helena had something else to recommend it: by all accounts, the weather was invariably clear, a matter of critical concern for astronomical observations.

3.3 *Winchester Street, London, where both Edmond Halley, the astronomer, and his father lived. As a young man, Halley made astronomical observations from these rooftops. This early nineteenth century print shows a street probably very little changed from Halley's time. (Ann Ronan Picture Library.)*

3.4 *The frontispiece of Halley's Chart of the Southern Skies. The dots representing the stars are connected with fanciful constellation forms. Around the periphery are constellations known to northern hemisphere astronomers, while toward the centre of the chart are the constellations that can be seen only from the southern sky. (Ann Ronan Picture Library.)*

Halley's new friends at the Royal Society wrote a letter to the government in support of the proposed expedition, which soon found favour with King Charles II. But the King was only the titular monarch of far-off St Helena; in reality, the little island was a fiefdom of the powerful East India Company. Charles II was, however, not without influence, and after he had written to the Company's directors, they volunteered to provide passage for Halley and a scientific companion. Halley's father, 'willing to gratify [Edmond's] Curiosity', provided him with a huge allowance, which more than covered the cost of the best observing equipment then available and whatever other expenses he might incur.

In November 1676, Halley set sail on the *Unity,* beginning a journey that would take three months and cover nearly ten thousand kilometres of ocean. His destination was an island so remote that, 130 years later, the British would deem it the only prison secure enough to contain the captive Emperor Napoleon. For almost two centuries, European mariners had been sailing the southern seas, mapping every coast line they could lay their eyes on, and not one of them had accurately mapped the very different constellations above them. Halley's self-imposed assignment was to bring back half the sky. He had just turned twenty-one.

Travellers' tales notwithstanding, the weather on St Helena was rotten. Halley would wait for weeks for an hour's window on the stars. *English* weather was better than this. At least in England, there were other things to do in inclement weather. Halley had exiled himself to little more than a rock in the middle of a vast ocean. And he had more than clouds to worry about. The Governor of St Helena was apparently a madman who grew to hate Edmond Halley. Eventually, the Governor's behaviour became so bizarre that he was finally recalled and dismissed, but not before Halley himself was ready to return. Despite the difficulties, his awful, trying year on St Helena was well spent. He came home with the first map of the southern skies, and much more. He had discovered stars and nebulae that European astronomers had not known about. While on St Helena, he had observed a transit of the planet Mercury across the face of the Sun, later significant in the determination of the Sun's distance from the Earth. He found that there was no Pole Star in the southern skies, and that accurate time-keeping on St Helena with a pendulum clock calibrated in England was impossible unless you shortened the pendulum. (Although Halley didn't know it at the time, this is because the spin of the Earth creates a small centrifugal force that slightly counteracts gravity at the equator, but weakens with increasing latitude.)

His catalogue of the southern skies was presented to the

Royal Society by Robert Hooke, a neurotic polymath who was the first person to see Jupiter's Great Red Spot through the telescope, and was the first person to see a living cell through the microscope. (He was also the first person to use the word 'cell' in its biological context.) Hooke made lasting contributions to physics, astronomy, biology, and engineering. The Fellows of the Royal Society were quick to appreciate Halley's achievement, but as far as Oxford was concerned, Halley was just another drop-out. They would not let him return to take his degree, because he had departed for St Helena before fulfilling his residency requirements at Oxford – so serious a breach of regulations that nothing short of a royal decree could remedy the circumstance. He appealed once more to Charles II and, once more, Charles signed a letter on his behalf, this time requesting that he be granted the degree of Master of Arts, 'without any condition of performing any previous or subsequent conditions of the same'. The Vice-Chancellor of Oxford complied. At about the same time he was granted his degree, Halley was elected a Fellow of the Royal Society, a considerable distinction for so young a man.

Halley soon turned his attention to an intensifying controversy involving Hooke, Flamsteed and Johannes Hevelius of Poland, the pre-eminent observational astronomer of the time. Hooke and Flamsteed swore by the recently invented telescopic sight, which, when mounted on a measuring instrument, improved the precision with which the relative position of the stars could be determined. But the older Hevelius rejected the new technology and persisted in using an open sight, without optics, a mechanism as simple as the sight on a rifle. Appalled, Hooke and Flamsteed conducted an increasingly shrill campaign against Hevelius, asserting to everyone who would listen that Hevelius' observations were untrustworthy. It was an unlikely topic for a vitriolic controversy. Hevelius, quite reasonably, began to feel beleaguered. He too was a Fellow of the Royal Society, but he lived in Danzig, far from the pubs and parlours in which Hooke and Flamsteed were pursuing their curious assault. He appealed to the Fellows, writing that every scientist should be permitted to serve astronomy as he believed best; that the results should be tested and judged solely on their merits; that it was unseemly for scientists to disparage the work of a colleague in the absence of supporting evidence. He had been observing with an open sight for more than thirty years, and if it had been good enough for Tycho Brahe it was good enough for him. The Royal Society responded by officially affirming its confidence in Hevelius and challenging Hooke and Flamsteed to document their case, or desist. They would do neither. The parties had reached an impasse.

3.5 *A sextant for measuring the positions of the stars being attended by Johannes Hevelius and his wife. This was the sort of technology that Halley came to Danzig to check out.*

3.6 *Johannes Hevelius of Danzig was also an observer of comets. Here, from his* Cometographia *(1688), are a variety of cometary forms between 1577 and 1652. The mottlings in the cometary head are caused by the telescopes of the time.*

3.7 *Comet Arend-Roland (1957), famous for its sunward spike. Compare with Hevelius' depiction of the comet of 1590 – the second comet from the top in the above figure. (Photographed through the University of Michigan telescope by F.D. Miller, 24 April 1597. Courtesy, NASA.)*

With his tact, integrity, and genius for observation – and his father's money – there could be no one better suited than Halley to make the journey to test Hevelius' methods. Halley sent Hevelius a copy of his southern star catalogue with a request for an invitation to visit. Hevelius agreed, and Halley arrived in Danzig in May of 1679. For ten nights, they observed together, and Halley was soon convinced that Hevelius, with the obsolete open sight, consistently achieved better results than Flamsteed and Hooke with state-of-the-art equipment. He immediately wrote to Flamsteed, describing his painstaking, systematic efforts to find flaws in Hevelius' observations: 'Verily I have seen the same distance repeated several times without any fallacy,' he informed his mentor, who was doubtless anxious to learn otherwise, '...so that I dare no more doubt his [Hevelius'] Veracitye.'

Despite Halley's testimonial, Flamsteed and Hooke lacked the capacity to apologise, or even to relent. The 'open sight' controversy did not die until Hevelius himself died, eight years later. Halley's conduct had been a model of fairness and candour, risking the displeasure of powerful friends in the interest of truth.

Although the names of many of the founders of modern science appeared on the Fellowship roster of the Royal Society in 1680, science was hardly widely accepted as the favoured approach to an understanding of nature. Gibbon mentions, as a perfect example of a world view in transition, an astronomer who 'was forced to allow that the tail, though not the head [of the comet of 1680] was a sign of the wrath of God'. Even Gottfried Kirch, the German astronomer who discovered the Great Comet of that year, was convinced of the supernatural nature of comets.

> I have read through many books on comets, heathen and Christian, religious and secular, Lutheran and Catholic, and they all declare comets to be signs of God's wrath... There are some that oppose the belief but they are not very important.

Halley also saw the Great Comet of 1680, but his response was different from Kirch's. On board a ferry in the English Channel, somewhere between Dover and Calais, Halley glanced up as the clouds broke and saw the splendour of the comet. As soon as he reached France, Halley hurried to the Paris Observatory to confer with its director.

Jean-Dominique Cassini, discoverer of the major gap in Saturn's rings as well as four of the planet's moons, received the young author of *A Catalogue of the Southern Stars* with lavish hospitality. Cassini entertained Halley, introduced him to friends and colleagues, gave him unrestricted use of the Observatory's equipment and library, and, most significantly, he presented Halley with an idea. In a letter to Hooke, dated May 1681, Halley wrote:

> Monsieur Cassini did me the favour to give me his booke of ye Comett Just as I was goeing out of towne; he, besides the Observations thereof, wch. he made till the 18 of March new stile, has given a theory of its Motion wch. is, that this Comet was the same with that that appeared to Tycho Anno 1577, that it performes its revolution in a great Circle including the earth...

Halley recounts the details of three cometary apparitions, adding:

> ...this is the sume of his Hypothesis and he says it will answer exactly enough to the Motions of the two Comets as likewise to that of Aprill 1665; I know you will with difficulty Embrace this Notion of his, but at the same tyme tis very remarkable that 3 Cometts should soe exactly trace the same path in the Heavens and with the same degrees of velocity...

No one had yet determined the orbit of a comet, but Cassini had noticed that three comets had come from the same part of the sky with similar speeds, and made the daring proposal – unprecedented in the scientific literature, and explicit only in the folk tradition of the Bantu-Kavirondo people of Africa – that the same comet was returning to Earth in widely separated times.

3.9 *More cometary forms by Hevelius. Note the comets with disrupted tails; the bottom three are called 'monstriferous comets', an echo of the once prevalent cometary mysticism.*

Some of the Italians called Pythagoreans say that the comet is one of the planets, but that it appears at great intervals of time and only rises a little above the horizon.

Aristotle,
Meteorology, Book 1,
Chapter 6.

Although not explicit, this seems also to imply the periodic return of the comets.

Halley informs Hooke that his preliminary attempts to sketch out a path for the comet of 1680, based on Cassini's summary of its apparent movement across the sky, have been unsuccessful; at some future time he hopes to try again. And in the final paragraph, as an aside, Halley lays the groundwork for the science of actuarial statistics. After summarizing the comparative statistics for Paris and London of births, marriages, deaths, and population density – the last facilitated by Halley himself pacing off the dimensions of the city of Paris on foot – he concludes:

> … it will from hence follow, supposing it alwaies the same, that one half of mankinde dies unmarried, and that it is necessary for each married Couple to have 4 Children one with another to keep mankinde at a stand…

It was ten years before he would develop this idea further, and even longer before he returned to the subject of comets. Now, he was off to Italy for six months. When he returned home to England, he made his first great discovery.

Her name was Mary Tooke. She was the daughter of the Auditor of the Exchequer. A contemporary memoir describes her as 'an agreeable young Gentlewoman; and a Person of real merit'. Another tribute portrays her as 'a young lady equally amiable for the grace of her person and the qualities of her mind'. They were married less than three months after his return from the Continent, at St James, a church notorious or compassionate – depending on your point of view – for choosing elopements as its speciality. There does not seem to have been any opposition to the marriage – nor was Mary pregnant. Perhaps they chose St James because they did not wish to waste another hour. Their marriage and their love were to last until Mary's death, nearly fifty-five years later. The few surviving references to them together make clear a deep and enduring happiness. In 1682, toward the end of their first summer together, Edmond, and possibly Mary, witnessed another comet, unimpressive compared to the Great Comet of 1680. He made a few notes to record what he saw. It was his only look at the comet that would one day bear his name.

Years after the death of Halley's mother, Edmond senior had entered into an unfortunate second marriage. There was talk about his new wife's extravagance, and her apparent disregard for her husband and his son. The contrast between the marriages of father and son must have pained them both. On the morning of 5 March 1684, the senior Halley complained that his shoes were too tight, and a nephew offered to cut the linings out of the toes. This seemed to help, and Edmond senior told his wife that he was going out; he would return by

3.10 *Halley's observations of the comet that would bear his name. On the left-hand page are his observations of 4 September 1682, and the right-hand page contains notes on the parabola, one of several candidates for the orbit of comets. (These two pages were reproduced in Arthur Stanley Eddington's article 'Halley's Observations on Halley's Comet, 1682', Nature, Volume 83, page 373, 1910.)*

nightfall. A contemporary broadside recounts what happened next:

> ...when Night came she accordingly expected him, but not returning she was very much concerned at it, and the next day made all possible Enquiry, but after several days not hearing of him, published his Absence in the News Book. From Wednesday the 5th of March, to the 14th of April, notwithstanding all Endeavours, and the strictest Search that could be made, they received no Account where he was, or where he had been. But on Monday last he was found by a River side at Temple-Farm in Strow'd Parish near Rochester on this manner. A poor Boy walking by the Water-side upon some Occasion spied the body of a Man dead and Stript, with only his Shoes and Stockings on, upon which he presently made a discovery of it to some others, which coming to the knowledge of a Gentleman, who had read the advertisement in the Gazet, he immediately came up to London, and acquainted Mrs Halley with it, withal, telling her, that what he had done, was not for the sake of the Reward, but upon Principles more Honourable and Christian, for as to the mony, he desired to make no advantage of it, but that it might be given intirely to the poor Boy; who found him and justly deserved it.

The same nephew who had adjusted Halley's shoes was dispatched by Mrs Halley to identify the body. It must have been a gruesome task: the face had been obliterated. The broadside continues:

> It was concluded by all, that he had not been in the River ever since he was missing, for if he had, his Body would have been more Corrupted. The Gentleman knew him by his Shooes and Stockings, they being the same Shooes he had cut the Lineing out of, and on one leg he had four Stockings, and on the other three and a Sear-cloath.* The Coroner sat upon him, & the Inquest brought him in Murthere'd.

The death of Edmond Halley senior remains a mystery. Unfortunately, no Sherlock Holmes was at hand to sift through the redundant footwear and deduce Halley's whereabouts during the five weeks previous. Two-and-a-half centuries after the event, Eugene Fairfield MacPike, the foremost Halley scholar, rendered a verdict of suicide, claiming that 'the evidence with which we are supplied appears to point to mental aberration'. Perhaps; but the evidence just as well points to murder.

An additional mystery revolves around the younger Halley's response to his father's fate. He is not mentioned in the broadside story, playing no apparent role in the search for his father, the identification of the body, or the Coroner's

*A cloth imbued with wax, and used as a bandage.

inquest. His whole life is a case study of a massive, almost compulsive curiosity. And yet we find no record that he made any attempt to resolve the mystery of his father's death – the man who had so generously encouraged and supported his intellectual development.

In an unseemly coda to these unhappy events, the 'Gentleman' who had brought the second Mrs Halley news of her husband's corpse sued for non-payment of the reward of £100 despite his protestations 'that he desired to make no advantage of it'. The case was heard by a Judge Jeffreys, who was infamous for the frequency and variety of the crimes he committed against the defendants who stood before him – harassment and extortion being the mildest among them. In this particular case, Judge Jeffreys seems to have taken the high road, awarding the 'Gentleman' only £20, and ordering Mrs Halley to pay the remaining £80 to the hapless 'poor Boy' who had discovered the body. A decade later, Edmond took his spendthrift stepmother to court in what was described as an effort to defend his inheritance.

At about the time that his father vanished, Halley had been trying to develop a deeper understanding of planetary motion. Kepler had pointed out that there is a precise proportion between the period it takes for a planet to go once around the Sun (its year) and the distance of the planet from the Sun. The year on Mercury, close to the Sun, is only 88 Earth days; the year on Saturn, far from the Sun, is almost thirty Earth years. The period of a planet in the outer solar system is longer than an Earth year, not only because it has a bigger orbit to traverse, but also because it is moving more slowly. Why? Halley and several others had a notion that the planets move as they do because of a balance of two forces – one directed outward from the Sun, and provided by a planet's own velocity, and the other directed inward, but provided by a previously undiscovered gravitational force from the Sun. It was clear that the force had to decline with distance, so that far-off planets could move slowly and still balance the force of gravity. But how fast must the gravitational force diminish as you move away from the Sun in order to account for the observed planetary motion? By intuition – or false analogy with the propagation of light, Halley and his colleagues suggested that gravity was an inverse square law. Move the planet twice as far from the Sun, and the force diminishes to a quarter its original strength; three times further away, and the force decreases to one-ninth; and so on. The law of gravity, whatever it might happen to be, dominates the heavens. This problem is central to our understanding of nature.

Halley, Hooke and Christopher Wren (1632-1723) – an

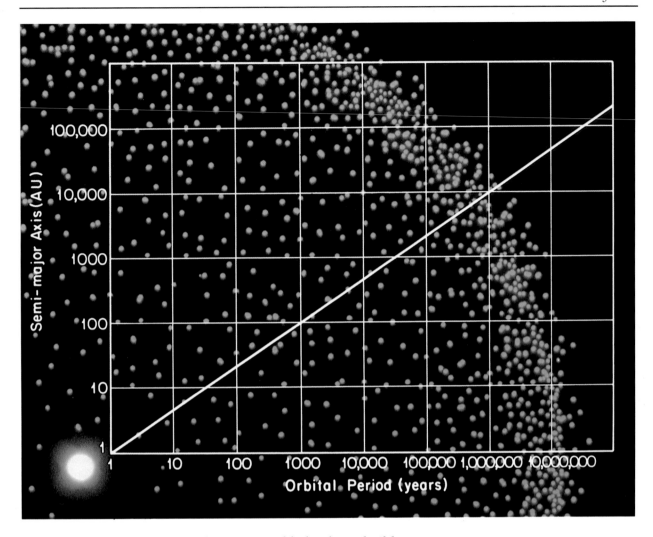

astronomer who turned to architecture and helped to rebuild London after the Great Fire, considered the challenge of proving the proposed inverse square law at a meeting of the Royal Society in January 1684. Hooke boasted that he had already done so, but declined to produce the proof. Wren must have had his doubts, because he offered a reward of any book that did not cost more than forty shillings to anyone who could provide a proof before two months had passed. Hooke maintained that he had found the solution, but was delaying showing it to others so all could appreciate the difficulty and significance of Hooke's own work. In fact, Wren was in no danger from this quarter of losing his forty shillings.

Months passed and the challenge remained unanswered, so Halley made up his mind to pay a visit to Trinity College, Cambridge, where there lived a man who might be equal to the task. This scholar had been judged a prodigy, primarily on the basis of his remarkable work on the nature of light and colours. But that had been many years before, and he had

3.11 *The time it takes for a planet or comet to orbit the Sun once increases the further from the Sun it is, according to a particular law of nature described by the diagonal straight line. The orbital period is given in Earth years and the distance from the Sun is given in Astronomical Units. At lower left we see that a body one AU from the Sun takes one year to go around the Sun – the case for the Earth. At a distance of 100 AU from the Sun, a comet would take a thousand years to orbit; and in the outer parts of the Oort Cloud of comets (see Chapter 11), indicated by the blue dots, a comet may take millions of years to go once around the Sun. This relationship, which was central to the work of Newton and Halley, is called Kepler's Third Law. (Diagram by Jon Lomberg/BPS.)*

squandered his genius during most of the interim in the pursuit of alchemical recipes and in strident attacks on the character of Athanasius, a theologian of the early Christian Church who did much to establish the orthodox doctrine of the Trinity. He believed Athanasius guilty of deliberate distortion and historical fraud. Unable to maintain normal relationships with anyone, especially women, this scholar was given to fits of paranoia and depression. Moreover, he was afflicted with an apparent inability to complete anything. Still, he was known to be a brilliant mathematician, and Cambridge was not so far from Halley's home in London. So, on an August morning in 1684, he set out to visit Isaac Newton.

Halley's encounter with Newton (1642-1727) in August of 1684 was a watershed for Halley, for Newton, for science, and, in ways so numerous as to be incalculable, for the fate of the world. The mathematician Abraham De Moivre took down Newton's version, related many years later, of this momentous encounter:

> ...the D^r [Halley] asked him what he thought the Curve would be that would be described by the Planets supposing the force of attraction towards the Sun to be reciprocal to the square of their distance from it. S^r [Newton] replied immediately that it would be an Ellipsis, the Doctor struck with joy & amazement asked him how he knew it, why saith he I have calculated it, whereupon Dr Halley asked him for his calculation without any farther delay, Sr Isaac looked among his papers but could not find it, but he promised to renew it, & then send it to him...

This promise might well have had a familiar and, therefore, hollow ring. But where Hooke prevaricated, Newton delivered. In November, a copy of Newton's *De motu corporum in gyrum* ('On the motion of bodies in orbits' – scholarly works were still being written in Latin) was delivered by hand to Halley. Only nine pages long, it contained the proof that the inverse square law implied all three of Kepler's Laws, as well as the seeds of a broad new science of dynamics. Halley instantly recognized what Newton had accomplished. Rushing back to Cambridge, he successfully extracted from Newton a promise to expand his ideas into a book, and to do it quickly.

Halley's first visit to Newton had awakened the latter from a kind of mystical trance. Now, Newton was wide awake to the point of sleeplessness, obsessed with this new challenge, unable to eat or to think of anything else. For the next year-and-a-half he would live as a recluse, in monomaniacal pursuit of gravity and planetary motion.

Meanwhile, Halley was back in London, busily trying to get

himself demoted. The Royal Society had operated since its inception in the early 1660s as a kind of club for gentlemen with scientific inclinations. But by 1685, the scientific revolution had assumed such dimensions that the voluntary services of its Fellows were inadequate to the task at hand. What was needed was a paid full-time secretary, someone who would deal with the growing correspondence, arrange the details of the meetings, and edit the *Philosophical Transactions*. Halley correctly understood that this position would afford him an ideal exposure to everything that was happening in science. He was elected to the job on the second ballot, early in 1686. But his salary from the Royal Society meant that he would have to relinquish his Fellowship, sit at the lower end of the conference table, and be denied the high honour of wearing a wig.

Cheerfully, Halley turned his voracious curiosity to the feast at hand – a banquet of geology, geography, biology, medicine, botany, meteorology, mathematics, and, of course, astronomy. He played a significant role in transforming the Royal Society from a club to the world's central clearinghouse for scientific ideas, and at the same time managed to publish many original papers of his own.

As Newton neared the completion of his masterwork, Halley approached the Royal Society with the proposition that they publish it. Since everyone agreed that the book would be important, the Society would, under normal circumstances, have been only too happy to pay for its publication. However, it had unwisely exhausted all its publication funds on another book. This was the long awaited *History of Fish*, which, unaccountably, had failed to find its anticipated audience. So Halley decided that he would pay for the publication of Newton's book with his own money. Oh yes, and the matter of his salary... This, too, would be a problem for the Society. Would Halley mind terribly taking the wages that had cost him his wig in remaindered copies of the *History of Fish?* Halley graciously accepted, and lugged home seventy-five copies of the *History of Fish*. By this time, in part because of his stepmother's incursions into his father's estate, he was no longer wealthy, but he chose not to make an issue of it.

The level of expectation surrounding Newton's forthcoming work was more than Robert Hooke could bear. He resurrected the old pretence that the inverse square law belonged to him. He would be satisfied with nothing less than an acknowledgement of his priority in the preface of Newton's book. Halley, who had already taken on the responsibilities of agent, editor, publisher, and proofreader for the book, now assumed yet another role: psychotherapist to the author. Fearing that Newton would hear of Hooke's charges from a less thoughtful source, Halley wrote him a letter. It begins with expressions of admiration and gratitude, but you can sense Halley's anxiety as he finally gets to the point:

> There is one thing more that I ought to informe you of, viz, that Mr Hook has some pretensions uipon the invention of ye rule of the decrease of Gravity, being reciprocally as the squares of the distances from the Centre. He sais you had the notion from him...

Newton's initial reaction was restrained, but the more he thought of it, the more enraged he became. He would withdraw the third volume of the work rather than become enmired in odious controversy with Hooke. But Book III was critical. Richard S. Westfall, one of Newton's biographers, writes of Book III:

> In a word, it proposed a new ideal of a quantitative science, based on the principle of [gravitational] attraction, which would account, not only for the gross phenomena of nature, but also for the minor deviations of the gross phenomena from their ideal patterns. Against the background of inherited natural philosophy, this was a conception no less revolutionary than the idea of universal gravitation itself.*

Book III also contained Newton's monumental work on comets. He had painstakingly collected observations of the comet of 1680, acquired from a wide range of locales – including London, Avignon, Rome, Boston, the island of Jamaica, Padua, Nuremburg, and the banks of the Patuxent River in Maryland. (Even then, worldwide co-operation was essential to understand the comets.) He showed that together they defined a highly eccentric orbit, almost a parabola (see Figure 3.12). Newton noted from examination of the history of comets that they are seen much more often in the part of the sky near the Sun than opposite the Sun, and understood this to be evidence that comets generally – not just the comet of 1680 – are in orbit about the Sun, and brighten when they are nearest to it. Tycho had shown that the comets move among the planets; now Newton, with Halley's prodding, had demonstrated that they have the same kinds of orbits (conic sections, see Figure 3.12) as the planets.

'The comets shine by the Sun's light, which they reflect', Newton wrote. 'Their tails... must be due either to the Sun's light reflected by a smoke arising from them, and dispersing itself through [space], or to the light of their own heads... The bodies of comets must be hid under their atmospheres.'

To Halley, it was unthinkable that these vital contributions might be lost to the world, a casualty of Newton's pique over Hooke's vanity and posturing. He reassured Newton that no

*Richard S. Westfall, *Never at Rest*, Cambridge University Press (1980).

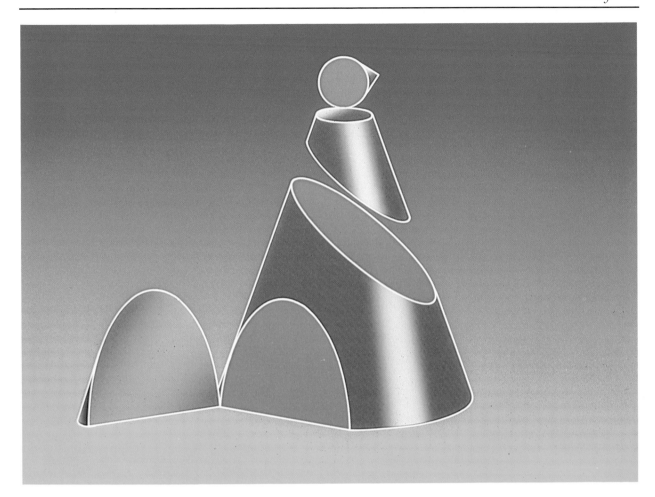

one took Hooke's claims seriously, and he told Newton the story of Wren's wager, back in 1684. He insisted that without Book III, the work would appeal only to mathematicians. He argued that Hooke's demands had been exaggerated by others. He cajoled, he flattered. Finally Newton relented, and allowed the publication of the entire work.

And this is how the *Philosophiae Naturalis Principia Mathematica*, ('The Mathematical Principles of Natural Philosophy'), the central testament of modern science, the keystone of our present understanding of stars, planets, comets, and much more, came to be. In the first edition of the *Principia*, as it has come to be known, printed in July of 1687, a poem of worshipful tribute presented to Newton by Halley appeared as preface. Its last lines read:

> Lend your sweet voice to warble Newton's praise,
> Who searcht out truth thro' all her mystic maze,
> Newton, by every fav'ring muse inspir'd,
> With all Apollo's radiations fir'd;
> Newton, that reach'd th' insuperable line,
> The nice barrier 'twixt human and divine.

3.12 *The conic sections. When a cone is sliced, or sectioned, at various angles, a variety of shapes are produced which collectively are called conic sections. Here, the exterior of the cone is shown in grey and three of the conic sections in blue. If the four pieces shown here were put together, the cone would be reassembled. The top of the cone is called the apex, and the flat bottom on which it stands is called the base, you produce a circle. If you cut at an angle, you produce a stretched-out curve, a little like an oval, called an ellipse. If you slice perpendicular to the base, the surface you expose is a hyperbola – unlike the circle and the ellipse, the hyperbola does not curve back upon itself. There is one other curve, not shown here, which lies at the very transition between an ellipse and a hyperbola; it is called a parabola. These pretty conic sections were first described by Apollonius. It is astonishing that planets and comets know to move about the Sun along such paths, and that if you throw a rock or launch a ballistic missile up into the air, the trajectory will follow a parabola. Newton showed that the inverse square law of gravitation forces bodies to move along conic sections. (Diagram by Jason LeBel/BPS.)*

But for him, in all human probability, the work would not have been thought of, nor when thought of written, nor when written printed.

Augustus De Morgan
(1806-1871) on Halley's
contribution to Newton's
Principia.

3.13 *Halley's interests also extended to questions of history and archaeology, as is shown by these papers published in the* Philosophical Transactions *on Julius Caesar's conquest of Britain and the ancient Near Eastern city of Palmyra. (Courtesy, The Royal Society of London.)*

(495)

A DISCOURSE *tending to prove at what* Time *and* Place, Julius Cefar *made his firft Defcent upon* Britain: *Read before the* Royal Society *by* E. HALLEY.

Though *Chronological* and *Hiftorical* Matters, may not feem fo properly the Subject of thefe Tracts, yet there having, in one of the late Meetings of the *Royal Society*, been fome Difcourfe about the Place where *Julius Cefar* Landed in *Britain*, and it having been required of me to fhew the Reafons why I concluded it to have been in the *Downs*; in doing thereof, I have had the good fortune fo far to pleafe thofe worthy Patrons of Learning I have the honour to ferve, that they thought fit to command it to be inferted in the *Philofophical Tranf-actions*, as an inftance of the great Ufe of *Aftronomical Computation* for fixing and afcertaining the Times of memorable Actions, when omitted or not duly delivered by the Hiftorian.

1. The Authors that mention this Expedition with any Circumftances, are *Cefar* in his *Commentaries lib.* 4, and *Dion Caffius* in *lib.* 39; *Livies* account being loft, in whofe 105th Book, might poffibly have been found the ftory more at Large. It is certain that this Expeditipn of *Cefars*, was in the Year of the *Confulate* of *Pompey*, and *Craffus*, which was in the Year of Rome 699. or the 55th before the ufual Era of Chrift: and as to the time of the year, *Cefar* fays that *Exigua parte aftatis reliqua*, he came over only with two Legions, *viz.* the 7th and 10th, and all Foot, in about 80 Sail of Mer-chant

III. *Some Account of the Ancient State of the City of* Palmyra, *with fhort Remarks upon the Infcriptions found there. By* E. Halley.

THE City of *Tadmor*, whofe Remains in Ruines do with fo much evidence demonftrate the once happy Condition thereof, feems very well to be proved to be the fame City which *Solomon* the Great King of *Ifrael* is faid to have founded under that Name in the *Defart*, both in 1 King. 9. 18. and 2 Chron. 8. 16. in the Tranflation of which, the *Vulgar Latin Verfion*, faid to be that of St. *Jerom*, has it, *Condidit Palmyram in Deferto.* And *Jofephus* (in *lib.* 8. *Antiq. Jud.* wherein he treats of *Solomon* and his Acts) tells us, that he built a City in the Defart, and called it *Thadamora*; and the *Syrians* at this day (fays he) call it by the fame Name: but the *Greeks* name it *Palmyra.* The Name is there-fore Greek, and confequently has no relation to the Latin *Palma*, and feems rather derived from Παλμυὸς or Παλμυς,

The judgment of posterity is not far from Halley's. Among the book's incidental offerings are the invention of the calculus, and the theory of interplanetary spaceflight, as well as the central idea behind the intercontinental ballistic missile. Newton's *Principia* changed the world.

The year after Halley's consummate midwifery delivered Newton of the *Principia*, he and Mary became the parents of two daughters, Katherine and Margaret. At about the same time, Halley became intrigued with Hooke's suggestion that the Flood, as related in the Bible, could be explained by a change in the Earth's poles, the Near East slowly sliding under the equatorial bulge of ocean. Halley was familiar with the continuous observations, made over a period of centuries, of minuscule changes in the latitude of the German city of Nuremberg, and reasoned that since even tiny changes in latitude occur with glacial slowness, then the time between the Creation and the Flood must have been far longer than the period allowed in Genesis, in contradiction of a literal interpretation of the Bible. However, in both the Old Testament and the ancient Babylonian accounts of the Flood, events proceeded swiftly, and recovery took less than a year.

Halley tried to imagine how a quick inundation of the ancient Near East could be worked. Were a comet to approach the Earth too closely, he argued, the gravitational tides might sweep the oceans (or just the Persian Gulf) up over a sufficient area of the land to account for the events described in the Bible. Halley also held that comets might, from time to time, actually impact the Earth, with still more horrible consequences. He seems to have been the first scientist to ask what might happen if a comet were to pass very near the Earth, a question now central to many areas of science, as we shall see.

Halley's interests were now more eclectic than ever. He attempted to measure the size of the atom. He made valuable observations about magnetism, heat, air, plants, seashells, clocks, caviar, light, Roman history, aerodynamics, the habits of cuttlefish, and a method of keeping flounder alive for mid-winter retailing. (These last two efforts were, perhaps, inspired by all those copies of the *History of Fish*.) By his own testimony, we know that Halley used opium. He gave a discourse on his personal experience of the drug at a meeting of the Royal Society, and does not seem to have suffered much from the 'amotivational syndrome' sometimes associated with opium and other euphorics. He also invented, developed and tested the first practical diving bell. 'By this means,' he wrote, 'I have kept 3 men 1-¾ [hours] under the water in ten fathoms deep without any of the least inconvenience and in as perfect freedom to act as if they had been above.' It worked so well that Halley was able to form a salvage company – it prospered, and eventually went public.

That same year, Halley published a paper on the means of determining the Earth's distance from the Sun – by measuring the timing of Venus as, on rare occasions, it transits across the Sun's disc. Much later, in 1716, he would publish another paper, imploring the astronomical community to organize co-operative international expeditions during the next transits. The first expedition of Captain James Cook in HMS *Endeavour* was intended specifically to respond to Halley's plea to measure a transit of Venus from Tahiti on 3 June 1769; for this reason alone (there are others), Halley plays a role in the history of the exploration of the Earth. From observations of the transits of 1761 and 1769, and with Halley's method of calculation, the astronomical unit was found to be nearly 93 million miles (150 million kilometres), a hair's breadth from the figure we accept today. Halley gave us the scale of the solar system.

In 1691, Halley was being considered for the Savilian Professorship of Astronomy at Oxford. His confirmation required the approval of the Anglican Church, then as now headed by the Sovereign. But Halley was accused of something scandalous: '…being guilty of asserting the eternity of the world'. (By these standards, many scientists and all Hindus would be disqualified from the Savilian Professorship today.) His crime had been to consider the cause of the Biblical deluge. Halley's conjectures on religious issues would today be considered irrelevant to his qualifications for teaching astronomy; moreover, the views ascribed to him by the authorities of the Church were a distortion of his true views. He never doubted that the Earth had been formed, or even created, and his efforts assumed the truth of the Biblical timescales, or at least of the duration of the Flood.

Flamsteed – the Astronomer Royal, or chief astronomical advisor to the Crown – seems to have played a disgraceful part in the affair. He was so miffed that Halley had disagreed with some work he had done on the ocean tides, that overnight Flamsteed transformed himself from friend to arch enemy. In a letter to Newton, Flamsteed insisted that the appointment be stopped, because otherwise the youth of Oxford would be corrupted by Halley's 'lewdness'. Even Newton, who was legendary for his prudishness, could not take this charge seriously. He urged Flamsteed to reconcile his differences with Halley, but the Astronomer Royal was unwilling. Now, the accusations went, Halley was not only lewd, but also a plagiarist, a thief of ideas. Throughout, Halley refused to be provoked. He uttered not an angry word in reply. When the integrity of his work was attacked, he scrupulously confined

3.14 *Edmond Halley's diving bell, with two aquanauts aboard. Fresh barrels of air were sent down from a boat above. The explorer on the sea floor is tethered by a breathing tube to the diving bell. This technology was successfully tested, and bears some resemblance to the early days of manned spaceflight. (From W. Hooper,* Rational Recreations, *London, 1782. Ann Ronan Picture Library.)*

his rejoinders to the scientific merits of the argument.

Sadly, the campaign by Flamsteed and the Church was successful, and Halley was denied the Savilian Professorship. During the interrogation on his alleged heretical beliefs, Halley made no attempt to curry favour. His inquisitor, a Chaplain Bentley, must have been outraged. Some forty years later, he published a tract entitled *The Analyst, or a Discourse Addressed to an Infidel Mathematician*, widely understood to be a harangue against Halley.

However Bentley may have scolded the applicant, it is clear that Halley emerged from the experience in good spirits, and with his scientific convictions intact. He continued his investigations into the age of the Earth, this time using the salinity of seawater as a kind of clock that had begun ticking at the time the oceans were formed. He thought that regular measurement of the waters of the ocean would reveal increasing salinity as time went on. Rivers carry salt into the oceans at a rate that Halley crudely calculated. Extrapolating back to a time when seawater was fresh, Halley found that the world is much older than the Bible implies – not 6,000 years, but at least a hundred million years old. Halley's method cannot be used to determine the exact age of the Earth because seawater is saturated with salt. But it is a perfectly suitable way of arriving at a lower limit for that age. Moreover, it is a brilliant anticipation of a range of modern techniques for dating rocks (see Chapter 16), and gave heart to geologists and biologists of the next century, when they came upon evidence that the Earth and life are ancient beyond all human imagining. The age of the Earth – and the rest of the solar system – is now known, from a range of independent lines of evidence, to be a little over 4.5 billion years.

When he was thirty-nine, Halley began the work for which he is chiefly remembered. Newton had demonstrated that the comets, like the planets, moved in orbits that were conic sections (see Figure 3.12), but which conic section was a matter of dispute. Newton himself was of the opinion that comets moved along open parabolic orbits; Cassini preferred circles; and the same Bishop of Salisbury whom Halley tried not to offend with his first published paper was leaning toward the ellipse. In addition, there were also champions of the hyperbola. It was difficult to know who was right, because Earthbound observers with feeble telescopes could see comets only when they came close to the Sun. The little arc that their paths described during this briefest and swiftest part of their journey could be accounted for by almost any of the conic sections, although circles were more difficult for their supporters to justify.

Halley approached this challenge – reconstructing a comet's

3.15 *The cover of Hevelius' Cometographia, depicting three seventeenth-century scholars debating the comparative merits of various hypotheses on cometary motion, as indicated in the diagrams they display. While the savants contend, the real thing appears in the sky above – unnoticed by them, but trapped by their assistants on the observatory roof. (Courtesy, Library of Congress.)*

whereabouts during the time it is invisible – with the discipline of a great detective. He steeped himself in every piece of recorded testimony on the subject, every surviving declaration made by a string of witnesses that stretched back to Pliny and Seneca. Through calculus and gravitational theory, Newton had provided the detection technology which Halley mastered. And there was also an element of luck; it was Halley's good fortune to live in a century graced with an abundance of comets, and therefore a mass of recent, relatively accurate, evidence on their behaviour.

Where the comet of 1682 was concerned, Halley regarded Flamsteed as the most reliable observer, and wrote to Newton, asking him to obtain the observations from Flamsteed, because 'he will not deny it you, though I know he will me'. Newton did as he was asked. Halley then compared the orbital characteristics, or elements, of the comets of 1531, 1607, and 1682, and found many striking similarities (see box) – in the tilt, or inclination, of the cometary orbit to the zodiacal or ecliptic plane (see Chapter 2); in the distance of the comet from the Sun at perihelion; in the region of the sky in which perihelion occurs; and in the place where the comet's orbit crosses the zodiacal plane (called the node). These similarities were already enough to suggest that the same comet was being seen in three different apparitions. When Halley compared the dates of the apparitions he found something like a periodic return – just as Newtonian theory had promised if the comets were on elliptical orbits.* The case was almost broken.

But Halley worried that the differences in orbital elements from one apparition to another, while small, were much larger than could be ascribed to errors of observation. The longitude of perihelion, for example, varied by more than a degree in the sky, yet the measurements were accurate to a few minutes of arc. Furthermore, the interval between the 1531 and the 1607 apparitions was more than a year longer than the interval between the 1607 and the 1682 apparitions. Thus, in addition to the metronomic regularity of an isolated comet in elliptical orbit around the Sun, there must, Halley believed, be some other influence or force that perturbed the comet one way in one apparition, and another way in the next.

Newton had proposed that the variations in the periods of the comets were produced by the gravitational attraction of undiscovered comets. But Halley knew that Jupiter and Saturn were perturbed by one another, and thought it likely that a comet, of much lower mass than either of these giant planets, would be more perturbed by even moderately distant

Aristotle's opinion... that comets were nothing else than sublunary vapours or airy meteors... prevailed so far amongst the Greeks, that this sublimest part of astronomy lay altogether neglected; since none could think it worthwhile to observe, and to give an account of the wandering and uncertain paths of vapours floating in the aether.

Edmond Halley
Transactions of the Royal Society of London, Volume 24, Page 882, (1706).

*Cassini had suggested something similar, but, as we now know, for apparitions of quite different comets.

approaches to either planet than by close approaches to another comet. He made a crude estimate of the effects on the comet's motion of the gravity of Jupiter and Saturn, and found they fit well with the measured discrepancies. Halley concluded that the slight differences in the orbital elements of the comets of 1531, 1607, and 1682 had a ready explanation. These were visitations of the same comet, vulnerable, as all travellers are, to detours, delays, and conditions *en route*.

Halley's investigation of comets constituted an enormous undertaking, requiring the painstaking calculation of the orbits of twenty-four comets that had achieved perihelion passage between 1337 and 1698. He was struck, as were

ORBITAL PROPERTIES OF THE COMETS KNOWN TO HALLEY

THE ASTRONOMICAL ELEMENT OF THE MOTIONS, IN A PARABOLIC ORBIT, OF ALL THE COMETS HITHERTO OBTAINED

Passage of Perihelion, London Time				Longitude of		Inclination of Orbit	Distance from Sun at Perihelion
				Perihe-lion	Asc. Node.		
	d.	h.	m.	° ′	° ′	° ′	
1337, June	2	6	25	37 59	84 21	32 11	0.40666
1472, February	28	22	23	45 34	281 46	5 20	0.54273
1531, August	**24**	**21**	**18**	**301 39**	**49 25**	**17 56**	**0.58700**
1532, October	19	22	12	111 7	80 27	32 36	0.50910
1556, April	11	21	23	278 50	175 42	32 6	0.46390
1577, October	26	18	45	129 32	25 52	74 83	0.18342
1580, Nov.	28	15	00	109 6	18 57	64 40	0.59628
1585, Sept.	27	19	20	8 51	37 42	6 4	1.09358
1590, January	29	3	45	216 54	225 31	29 41	0.57661
1596, July	31	19	55	228 15	3.2 12	55 12	0.51293
1607, October	**16**	**3**	**50**	**302 16**	**50 21**	**17 2**	**0.58680**
1618, October	29	12	23	2 14	76 1	37 34	0.37975
1652, Nov.	2	15	40	28 19	88 10	79 28	0.84750
1661, January	16	23	41	115 59	82 30	32 36	0.44851
1664, Nov.	24	11	52	130 41	81 14	21 18	1.02576
1665, April	14	5	16	71 54	228 2	76 5	0.10649
1672, February	20	8	37	47 0	297 80	83 22	0.69739
1677, April	25	00	38	137 37	236 49	79 3	0.28059
1680, Dec.	8	00	6	262 40	272 2	60 56	0.00612
1682, Sept.	**4**	**7**	**39**	**302 53**	**51 16**	**17 56**	**0.58328**
1683, July	3	2	50	85 30	173 23	83 11	0.56020
1684, May	29	10	16	238 52	268 15	65 49	0.96015
1688, Sept.	6	14	33	77 0	350 35	31 22	0.32500
1698, October	8	16	57	270 51	267 44	11 46	0.69129

This table, taken from Halley's work, is the basis for his conclusion that the three apparitions here highlighted refer to the same comet. Dates are given in days, hours and minutes; angles in the sky in degrees and minutes of arc; and perihelion distance in Astronomical Units (the Earth is 1 AU from the Sun).

Aristotle and Seneca, by the random character of their inclinations:

> Their orbits are disposed in no certain order... They are not confined like the planets to the Zodiac, but... move indifferently, every way both retrograde and direct.*

In their distance from the Sun at perihelion passage, the comets Halley studied ranged from about one Astronomical Unit (AU) to less than 0.01 AU – this was the Great Comet of 1680, that nearly grazed the Sun. And at aphelion, he found the comets – including the comet of 1682 – ranged well beyond the orbit of Saturn, then the most distant planet known.

In 1705, Halley published the results of his 'immense labour' in a book entitled *A Synopsis of the Astronomy of Comets.* It was the first application of the laws of the universe, as revealed by Newton, by a scientist other than Newton to solve an astronomical mystery. This, in itself, was enough to guarantee Halley a place in the history of science. But he went much further, making a courageous leap that earns him a place in the larger history of civilization. For millennia, comets had been the almost exclusive property of the mystics, people who considered comets as portents, symbols, wraiths – but not *things*. Halley shattered their monopoly by beating them at their own game, a game that no scientist had ever played before: prophecy. He predicted that the comet seen in 1531, 1607 and 1682 would return more than fifty years in the future. And he did not hedge his bet. It would return, he stated flatly, at Christmas time in 1758 – from a particular part of the sky, with specific orbital elements. There is hardly a prophecy of the mystics that even strives for comparable precision. He was, for his time, remarkably free of nationalism, but this once he indulged himself in a bit of national chauvinism:

> Wherefore if it should return according to our prediction about the year 1758, impartial posterity will not refuse to acknowledge that this was first discovered by an Englishman.

In the spring of 1696, Newton was appointed Warden of the Royal Mint. It required him to live in the Tower of London, and his task was to prevent clipping or shaving of the coinage of the realm. Months later, Newton – who had just begun his study of comets – appointed Halley as Deputy Comptroller of the Mint at Chester. Under the Master and the Warden in the Tower, were Masters, Wardens, Comptrollers and the like in provincial facilities of the Mint. Halley spent two miserable years at Chester supervising the mechanized production of milled coins out of the old hand-made ones. When he and his

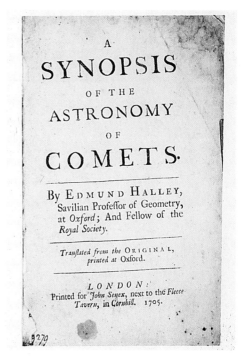

3.16 *Cover page of the 1705 work by Edmond Halley, in which the orbits of comets are calculated, and the periodic return is postulated of the comet which posterity has named after Halley.*

---

*That is, both anti-clockwise and clockwise as seen from a vantage point high above the North Pole.

local Warden discovered two clerks skimming precious metals for their own gain, they spoke out – unaware that their own superior, the Master of the Mint at Chester, was getting a cut from the clerks. An acrimonious feud followed and there were threats of a duel; but the Mints were closed in 1698, and Halley was free to return to London.

He arrived just in time for a more agreeable assignment. The twenty-six-year-old Czar of Russia, who would later be known as Peter the Great, had come to England to learn how to Westernize his nation. He was ensconced at a grand country home near the Deptford shipyards, where he performed manual labour in an effort to gain first-hand experience of the British genius for shipbuilding. He had hoped to spend time with Newton, but Newton sent Halley in his stead. By all accounts, the English astronomer and the Russian Czar became fast friends, sharing passions both for knowledge and for brandy. For the duration of Peter's stay in England, Halley remained his scientific advisor and drinking partner. One disputed, although contemporary, account depicts Halley raucously pushing the Czar of all the Russias through the streets of Deptford in a wheelbarrow in the dead of night. It was said that the two of them were inebriated, and that the escapade caused grave damage to the topiary hedges.

That same year, Mary Halley gave birth to a son named Edmond, and her husband began yet another career, described in a contemporary account replete with italics:

> In 1698 the King [William III] who had been inform'd of Mr Halley's ingenious Theory of the Magnetic Needle, was desireous the variation shou'd, for the Benefit of Navigation, be carefully observed, in diverse parts of the Atlantic Ocean; for which purpose His Majesty, the 19th Aug. 1698, appointed Mr Halley Commander of His Ship the Paramoor* Pink, with orders *to seek by observation the discovery of the Rule of the variation of the Compass,* and at the same time *to call at his Majestys Settlements in America, making som observations there, in order to the better laying down the Longitudes and Latitudes of those places, and to attempt a discovery of what Lands lay to the South of the Western Ocean.*

Seventy years before Cook set sail (in pursuit of a Halleyan objective), Halley was commanding the first marine scientific expedition to be commissioned by a British monarch. The *Paramour* sailed to Spain, the Canary Islands, Africa, Brazil and the West Indies, before an incipient mutiny on the part of Halley's second-in-command necessitated an unscheduled return to England. In the course of the Court Martial that ensued, it was revealed that the disgruntled lieutenant was something of a closet magnetic theorist, whose own writings

---

*Spelling conventions were less rigid in Halley's time than they are today.

on the subject had earlier been found wanting by the Royal Society. He had nursed his indignation, and when the landlubber Halley was commissioned as his captain, had become unhinged. Halley assumed the role of navigator and took great pride in the fact that he was able to bring his ship home without the loss of a single life.

Halley commanded two additional voyages on the *Paramour* headed north – one, amidst many perils, up the coast of South America to Trinidad, bound for Cape Cod. They found the seas off Nantucket too high, and were forced to sail on to Newfoundland. There, the *Paramour* took friendly fire from an English fishing fleet that mistook her for a pirate ship. Again, the Captain safely returned his ship to England, but this time he was denied the satisfaction of seeing every member of his crew at the homecoming. The cabin boy had been swept overboard and lost during a wild storm off the Canary Islands. For the rest of Halley's life, he was unable to speak of the boy's death without tears.

Halley fulfilled his royal mandate by publishing 'A *New and Correct* CHART *Shewing the* VARIATIONS *of the* COMPASS *in the* WESTERN AND SOUTHERN OCEANS *as observed in ye* YEAR 1700 *by his Maties* [Majesty's] *Command'*. The map contained a dotted-line convention that Halley devised to indicate points of equal variation in the Earth's magnetic field, the method that remains in use in magnetic maps today. Halley expanded his Atlantic Chart to a World Chart that remained in print, through many editions, for a century.

Halley's third and final voyage as captain of the *Paramour* in 1701, kept him closer to home. His purpose was to study the tides of the English Channel, although it has been suggested that his hidden agenda included a reconnaissance of the French coast on the eve of the War of the Spanish Succession. The following year, Queen Anne dispatched Halley on diplomatic missions to several European monarchs.

When Halley returned to England, he was, perhaps to his surprise, offered the Savilian Professorship at Oxford – but of geometry, not astronomy. Thirteen years after his vitriolic campaign against Halley's appointment to another Savilian chair, Flamsteed wrote a carping letter to a mutual friend, denying both Halley's suitability and his prospects for the chair, and complaining, 'He now talks, swears, and drinks brandy like a sea captain...' But by now, Halley was too well respected for Flamsteed's venom to do much harm, and in 1704 he was appointed.

His inaugural lecture was a loving tribute to the geometrical achievements of his colleagues. Newton, of course, was singled out for the most lavish praise. Halley dedicated much of his tenure as Savilian professor to a rediscovery of the

With the appearance of these magnetic charts and his already published work on the physics of wind, monsoons, and evaporation of seawater, Halley can truly be termed the founder of modern geophysics, a fact which was recognized by the Royal Society in 1957 during the International Geophysical Year, when it named its permanent scientific base in Antarctica Halley Bay.

Colin Ronan
Edmond Halley: Genius in Eclipse
(Doubleday, New York, 1969).

much it past above the Pleiades and how much under the Belt of Orion...' Halley could not help himself; he was desperate to know, and his desire was even greater than his celebrated kindness. He wanted to know everything about the meteor: its altitude, velocity, what sound it made, how big it was, of what it was made. We cannot help but feel privileged to know the answers to his questions (see Chapter 13); we cannot help but wish it were somehow possible to share them with him.

At the age of sixty-five, in an act of consummate optimism, Halley undertook an ambitious study of the eighteen-year cycle of solar eclipses. Halley, the inventor of the actuarial table, could not have been unaware of the improbability that he would live long enough to complete the project. He confounded the odds, and finished when he was eighty-four.

His productivity and his longevity were extraordinary in another respect. Physicists are sometimes said to be like mayflies, with only a brief creative period; indeed, a strikingly large fraction of major discoveries are made by physicists under the age of thirty-five. This is more true for theoretical than for experimental physics, and more true for physics than for astronomy. Perhaps, when past thirty, the mind loses some of its capacity to conceptualize on a grand scale. But in the last decades of Halley's life, he made major theoretical advances in our understanding of nature on its grandest scale: the universe. He discovered that the so-called 'fixed' stars actually moved with respect to one another. The discovery may have been stimulated by work on Flamsteed's book. It took another hundred years of instrumental development in astronomy before Halley's discovery of stellar proper motion could be confirmed. In another paper of his last years, he anticipated the discoveries of a much later time, arguing a limitless universe without a centre. Halley was guilty of believing in infinity.

Mary Halley died when he was eighty years old. Shortly after, Halley suffered a stroke and the loss of his son. Despite these blows, he continued making astronomical observations and attending scientific meetings until a few weeks before his death on 14 January 1742, at the age of eighty-six. His last words were a request for a glass of wine. He drank it while sitting in a chair. When the glass was empty, he died without a sound.

He had wished to lie next to Mary forever. His daughters engraved this tribute (translated from the original Latin) on their parents' tomb:

> Under this marble peacefully rests, with his beloved wife, Edmond Halley, LL.D., unquestionably the greatest astronomer of his age. But to conceive an adequate knowledge of the excellencies of this great man, the reader

He seems to be not one but all mankind's epitome.

Herbert Dingle,
The Halley Lecture
(Oxford University,
1956).

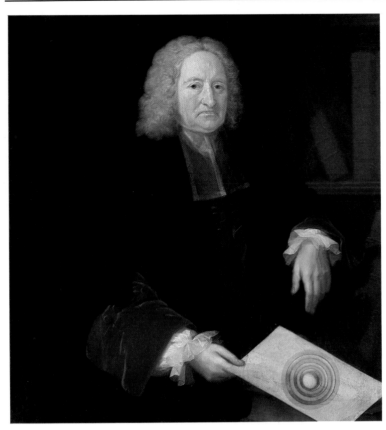

3.17 *A portrait of Edmond Halley at age eighty, just before the death of his wife, Mary. (Painting by Michael Dahl. Courtesy, The Royal Society of London.)*

must have recourse to his writings; in which almost all the sciences are in the most beautiful and perspicacious manner illustrated and improved. As when living, he was so highly esteemed by his countrymen, gratitude requires that his memory should be respected by posterity. To the memory of the best of parents their affectionate daughters have erected this monument, in the year 1742.

Edmond Halley was not just a man who discovered a comet. In fact, discovering a comet was one of the few scientific activities in which he did not engage.

There is a widely held belief that the price of deeply understanding the complexities of nature is paid out in the currency of alienation. Something like an inverse square law exists, it is said, between scientific genius and the capacity for love. If the life of Isaac Newton is pointed to as the most dramatic illustration of this theorem, then the life of his friend Edmond Halley, who more than anyone else made manifest Newton's genius, may be offered as its most inspiring exception.

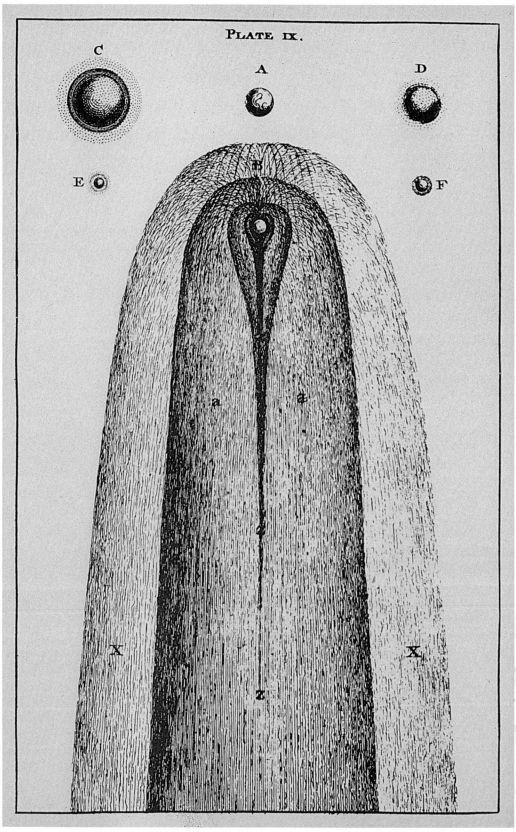

PLATE IX.

Chapter 4

THE TIME OF THE RETURN

Guardian and friend of the moon, O Earth, whom the comets
forget not,
Yea, in the measureless distance wheel round and again they
behold thee!

Samuel Taylor Coleridge,
Hymn to the Earth
(1834).

China invaded Tibet and Turkestan; French troops seized the Ohio Valley; Britain declared war on France; Germany defeated Austria, whereupon Austria defeated Germany; a Russian army occupied Germany; and an Indian revolution against the British army of occupation was ruthlessly put down. In such respects, the decade of the 1750s was almost indistinguishable from many others. But it was also a time of enlightenment. Diderot's *Encyclopaedia* was published in France and Samuel Johnson's *Dictionary* in England; Hume, Rousseau and Voltaire wrote seminal works, Bach died and Mozart was born. Lomonosov founded the University of Moscow. The Prussian Academy of Sciences and the first mental asylum in London both opened their doors, *Tristram Shandy* was being written, Hokusai was born in Tokyo, and a little-known Virginia surveyor named George Washington married a dowdy widow named Martha Custis.

In science, this was the decade that would end with the return of a comet – if Edmond Halley's prediction could be believed. In the first half of the decade, two extraordinary scientific works were published, each bearing on the nature of comets, and each presenting a view of the universe astonishingly in advance of their times.

Thomas Wright of Durham was an astronomer by temperament, although entirely self-taught. Born in the north of England in 1711, the son of a carpenter, he was prevented from continuing his early schooling 'by a very great impediment of speech'. He also seems to have been expelled from school on account of his behaviour. He described himself as 'very wild and much addicted [to] sport'. Following the practice of the time, at age thirteen he was apprenticed – to a

4.1 *(Opposite) A drawing of the nucleus, coma, and tail of the Great Comet of 1680, and the nuclei of four other comets, with the Earth (A) shown for scale. (From Thomas Wright's* An Original Theory of the Universe, *London, 1750.)*

4.2 *Thomas Wright of Durham as depicted in an engraving in* Gentlemen's Magazine, *January 1793. The portrait is circumscribed by a serpent. (Ann Ronan Picture Library.)*

4.3 *The orbit of a comet cuts through a solar system circumscribed, just outside the orbit of Saturn, by a serpent. (From Thomas Wright's* An Original Theory of the Universe, *London, 1750.)*

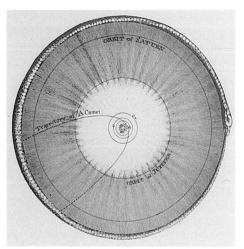

clockmaker, where he spent so much of his time poring over the astronomical literature that his father though him mad. In contrast to Edmond Halley's father, Wright's father apparently attempted to influence his course of study by burning his books. Shortly afterwards, Wright was dismissed from his apprenticeship amidst considerable scandal – he fell in love with a clergyman's daughter, but discovered his plans for a secret marriage 'prevented, and Miss lock'd up'; in his anguish, he took a passage to the West Indies, but was prevented from embarking by his outraged father.

With this promising start, he taught himself surveying and navigation, became a tutor to the children of the aristocracy, turned down a professorship at the Russian Imperial Academy in St Petersburg, and began writing books on astronomy. The most extraordinary of them, called *An Original Theory of the Universe*, was just that. Published in 1750, it is the first known statement of the true nature and geometry of the Milky Way – not a road of the gods, not divine milk splashed across the heavens, not an architectural support holding up the sky, but a flat disc of stars each like the Sun, all suspended in the ocean of space. From the time of Democritus, there had been a few people who had guessed that the Milky Way was made of individual stars too faint and distant to be seen individually; and this notion had been vindicated by Galileo with the first small telescope. So by Milton's time, it was possible for a poet to describe the Milky Way as a galaxy, 'powder'd with stars'. But the idea of the Milky Way as a flattened concentration of stars in which the Sun is embedded was first proposed by Wright. He even imagined that the stars revolved about some centre of the galaxy 'as do the planets around the Sun'.

While there are mystical elements in Wright's writings, and certainly not all that he proposed in his *Original Theory* has stood the test of time, his vision of the Milky Way is a landmark in the history of astronomy. The work is the more remarkable since its author had never acquired a formal education. This vision – a galaxy full of moving stars – turns out, as we shall see later, to be central to any understanding of the nature and origin of comets.

But comets were discussed in their own right in *An Original Theory.* Using his talents as draftsman and surveyor, Wright designed elegant diagrams of the solar system, with comets much in evidence. He delighted in showing the cometary orbits tabulated by Halley in their correct sizes and orientations (see Figures 4.3 and 4.4), providing many readers with a jolting first look at the comparatively small scales of the planetary orbits. In Figure 4.5, Wright attempted to show the relative sizes of the worlds then known. The massive planets,

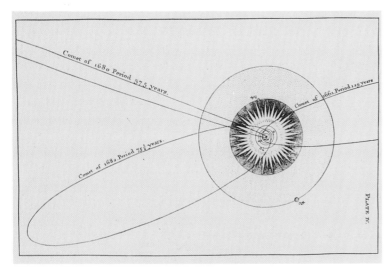

4.4 *Thomas Wright's elegant depiction of the solar system as known in his day. At the centre are the planets Mercury, Venus, Earth, and Mars – each represented by their astronomical symbol – and then, just outside the Sun's rays, the orbits of Jupiter and Saturn. Of the three comets shown, the Comet of 1680 is the one Newton had first computed the orbit of, and the Comet of 1682 is the comet Halley had predicted would return only a few years after the publication of this book. (From Thomas Wright's* An Original Theory of the Universe, *London, 1750.)*

Jupiter and Saturn, dominate the drawing. Their nine moons, all that were then known, and the inner planets – Mercury, Venus, the Earth and Mars, along with the Earth's moon – are shown as insignificant beside the two giant planets then known. Jupiter had already been observed to be cloud-covered, and Wright could not resist showing some land and ocean peeking through the overcast. He felt those worlds beckoning.

With this fine beginning, he attempted to draw the comets to scale as well, from the observations of comets available in his time. The result is shown in the frontispiece to this chapter. For scale, A represents the Earth; and C, D, E, and F, the nuclei of the comets of 1682, 1665, 1742, and 1744 respectively. In fact, what was being measured in Wright's day – and this is still largely true in our own – was not the nucleus of the comet, but the coma. The nucleus – Wright used the term – is the bright, solid object at the centre of the comet, the presumed source of the fine particles or gas that constitute the comet's tail. But the coma, the cloud of matter around the nucleus, shields it from our view. The almost total absence of detail in Wright's drawings of cometary 'nuclei' could have provided a hint that the nucleus is surrounded by the coma. The nucleus might be much smaller than the coma; it might even be too small for any detail to be visible, even if it were not enveloped in a shroud of matter. As the frontispiece indicates, the comas of comets in the vicinity of the Earth can be the same size as the Earth, or even larger.

Prominent in the frontispiece is Wright's depiction of the Great Comet of 1680, the very comet that Newton had, in *The Principia,* so brilliantly demonstrated was orbiting in a conic section and obeying the law of universal gravitation. This was also the comet that had helped to stimulate Halley's interest.

4.5 *Portraits of Jupiter and its four moons (top), Saturn and its five moons (bottom), and the planets Mercury, Venus, Earth, and Mars, with the Earth's Moon thrown in for good measure in the middle. The Cassini Division in the rings of Saturn is shown; the details on the surfaces of Jupiter and Saturn are wholly fanciful. (From Thomas Wright's* An Original Theory of the Universe, *London, 1750.)*

We do not know whether Wright merely drew what others had described in words, or whether he had access to drawings made at the telescope, but what is depicted here is close to the appearance of many comets (see Chapters 7 and 10).

Wright described the region labelled *aa* as 'the comet's natural atmosphere', the converging central lines representing 'denser matter'. *XX* is a representation of 'the inflamed atmosphere and tail dilated near the Sun'. The 'nucleus' of the comet of 1680, as we have already remarked, is merely the inner coma. Above it are three concentric shells of matter that we know from observations of subsequent comets are commonly ejected, in succession, from the nucleus. The traceries of fine lines in this outer coma have a curious appearance, as if fountains of material were pouring out into space on the sunlit side of the cometary nucleus.

To sum up the scale of the solar system, Wright produced, for his readers' amazement and delight, a three-part diagram. In Figure 4.4, the size of the Sun is set to scale with the orbit of Mercury. In Figure 4.5, to scale with the orbit of Mercury, are arcs representing, successively, the orbits of the remaining planets known in Wright's time – Mercury, Venus, the Earth, Mars, Jupiter and Saturn. In Figure 4.6 the entire planetary system is represented by a central dot, with a rosette of cometary orbits the only token of where the planets might lie.

Wright took the outer boundary of the solar system to be a little beyond the farthest reaches of the orbits of the most distant comets then known. He then drew the 'least possible distance' between the Sun and what was then considered to be the nearest star, Sirius. (We now know that both the outer boundaries of the solar system and the distance from the Sun to Sirius are vastly greater than was imagined in Wright's day.) He conjured up 'a new-created Mind or thinking Being, in a profound State of Ignorance', who sees both Sirius and the Sun from a remote vantage point, and then arrives in the solar system and observes the disposition of comets and planets. What would this being imagine is in orbit around Sirius? Wright cheerfully gives the answer: 'Why Planets such as ours'. And comets.

Wright's book probably influenced the perspective of astronomers and the future of astronomy in ways that can no longer be traced, but its most significant known influence occurred because of a review that appeared in the following year (1751) in a German magazine called *Achtes Jahr* ('Biquarterly'). One of its readers, living in the town of Konigsberg, was Immanuel Kant, a twenty-seven-year-old graduate student who was much taken with the scope and

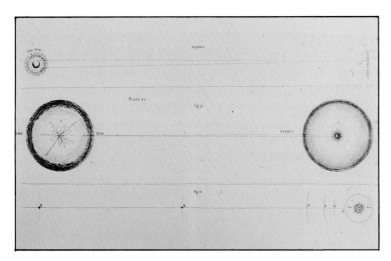

4.6 *The scale of the solar system, according to Thomas Wright. Here, the cometary rosette (left) is very reminiscent of the Bohr atom, while 'Syrius' is depicted as something at least very close to the eye of God. Compare with Figures 2-5 in Chapter 1. (From Thomas Wright's* An Original Theory of the Universe, *London, 1750.)*

elegance of Isaac Newton's work. He was later to become a towering figure in philosophy, but in the early 1750s, he was mainly interested in science. Four years later, aroused by Wright's vision of the universe, but still having read only the review of his book, Kant published the *General Natural History and Theory of the Heavens*. In it, he makes clear his debt to Wright. Kant's book was published four years before the expected return of Halley's Comet.

In their intellectual lives, Kant and Wright were kin. But in their personal lives, there were at least as many differences as similarities. Besides his native intelligence, Kant had few natural advantages. Unlike Wright, he was encouraged by his parents. In poor health his whole life, with a deformed chest and barely five feet tall, Kant subjected himself to a strict regime of regular exercise, mainly walks. His father was a saddle-maker. Where Wright had at least *planned* to run away with the vicar's daughter, Kant seems not to have had a close relationship with any woman besides his mother. While Wright was prepared to seek his fortune in far-off America, Kant never ventured more than a hundred kilometres from Konigsberg in his entire life. Where Wright precipitated scandals, Kant's behaviour was proper, even a little austere. Where Wright was thrown out of school, Kant was a highly successful student, admired by his teachers. Wright and Kant were different men, moulded by different cultures, but they were equally swept up in the great Newtonian vision of innumerable stars and innumerable worlds all moving in solemn obedience to a great universal law of gravitation – one that could reconstruct their positions in the distant past and predict their positions in the far future.

Kant's *Theory of the Heavens** blazed many trails. He accepted Wright's view of the Milky Way as a flattened volume of space bounded by two parallel planes and filled with stars. Kant was the first person to consider the origin and evolution of the galaxy, a central topic of modern astrophysics. And he made a daring additional leap – he supposed that the Milky Way was one of innumerable other galaxies, each filled with stars and, it may be, planets and life, a cosmic perspective not fully demonstrated until the 1920s. Wright had reached for such a vision, but had not grasped it. Kant correctly proposed that the spiral nebulae, such as M-31 in the constellation Andromeda (see Figure 4.8), were distant Milky Ways. That we live in a universe of galaxies, each composed of a multitude of suns, is perhaps the central revelation of modern astronomy.

There are many other delights to be gathered from the pages

---

*The full title was *Universal Natural History and Theory of the Heavens, Or An Essay on the Constitution and the Mechanical Origin of the Whole Universe Treated According to Newton's Principles* – an ambitious mouthful, but Kant delivered.

of the *Theory of the Heavens*, including the first statement that the solar system had been formed from a cloud of diffuse interstellar matter. The idea is known today as the Kant-Laplace hypothesis, and the cloud of matter is now called a solar nebula or, increasingly, an accretion disc. Naturally enough for someone writing on astronomy in the decade of the predicted return of Halley's Comet, Kant discussed comets:

> The most distinctive mark of the comets is their eccentricity... Their atmospheres and tails, which on their great approach to the Sun are spread out by its heat... [were] in ages of ignorance unusual objects of terror, [and were] regarded by the common people as foretelling imaginary fates... It is not possible to regard the comets as a peculiar species of heavenly bodies entirely distinct from the race of planets.

He pictured comets condensing from 'primitive matter in regions of space far away from the centre and which is feebly moved by [gravitational] attraction'. He imagined that, unlike the planets, the comets formed at all inclinations, a situation he described as 'lawless freedom'. 'Hence,' he said, 'the comets will come to us without restriction from all quarters.'

4.7 *A photograph, from our position embedded inside the Milky Way, of the centre of the galaxy beyond the nearby constellation Sagittarius, which provides the foreground stars. We see the great clouds of gas and dust, cutting diagonally through the picture and marking the plane of the galaxy. (Photograph by David Talent, courtesy, Association of Universities for Research in Astronomy Inc., Cerro Tololo Inter-American Observatory, Chile.)*

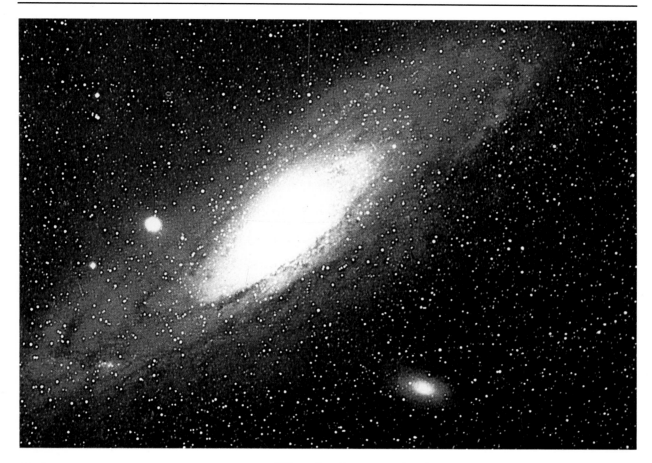

4.8 *The great galaxy M-31 in the constellation Andromeda, which supplies the foreground stars. This is the nearest spiral galaxy, like our own Milky Way. No individual stars in M-31 are resolved in this photograph, but it contains hundreds of billions of such stars. If this were a photograph of our own Milky Way, somehow obtained from far outside, the position of the Sun would be so far from the centre as to be off the edge of the picture. M-31 is shown accompanied by two small satellite galaxies. (Courtesy, Hale Observatory, Carnegie Institution of Washington and California Institute of Technology.)*

When he discusses the density and, in effect, chemistry of the comets, he gets off to a breathtakingly good start:

> The specific density of the matter out of which the comets arise is more remarkable than the magnitude of their masses. As they are formed in the furthest region of the universe, it is probable that the particles of which they are composed are of the lightest kind; and it cannot be doubted that this is the chief cause of the vapour heads and tails by which they are distinguished from other heavenly bodies.

But now a wrong turn:

> This dispersion of the matter of the comets into vapour cannot be attributed mainly to the action of the heat of the Sun; for some comets scarcely reach as near the Sun as the distance of the Earth's orbit; and many stop between the orbits of the Earth and Venus, and then turn back. If such a moderate degree of heat resolves and attenuates the matter on the surface of these bodies to this degree, must they not consist of the very lightest stuff, which can suffer more attenuation by heat than any other matter in the whole of nature?

Kant's problem is to find a material which turns from solid into vapour at the Earth's distance from the Sun. He had only to look out of his window, perhaps at the very moment he was writing these lines, to see a cloud of vapour rising off the frozen River Pregel, and he would have had his answer: ice – ordinary water ice. Not an exotic substance, at least for the Earth. No special celestial stuff. Just ice.

Of course it is unfair to Kant – whose vision on these matters was remarkable in a remarkable age – to criticize him with hindsight. We are the beneficiaries of the work of large numbers of capable scientists who have lived between his day and ours. But picture him, quill pen in hand, wondering about what this extraordinary 'lightest stuff' might be, trying to grasp the quintessential, when it was all around him. You want to reach across the centuries and give him a word of encouragement. Of course there may be a simpler explanation. Perhaps his study had no window. Perhaps he wrote the third chapter of the second part of the *Theory of the Heavens* in the summer.

Elsewhere in the book, he almost arrives at the right answer by a wholly unexpected route. He imagines that at some time in the past, the Earth itself was graced by a ring like the rings of Saturn – which he correctly believed to be composed of individually orbiting worldlets, perhaps made of ice. He exults at how lovely the sky would have been:

> A ring 'round the Earth! What a beautiful sight for those who were created to inhabit the Earth as a paradise! What a convenience for those on whom nature was designed to smile on all sides!

It strikes him that such a ring might explain the odd phrase in the book of Genesis, 'the water upon the firmament' – waters that in some sense were intrinsic to the heavens. Kant notes that this idea 'has already caused not a little trouble to commentators', that is, those struggling to reconcile physics and the Bible. Indeed, Thomas Aquinas had spent time on this very question in the *Summa Theologica*. Kant had a suggestion for the biblical exegetes:

> Might this ring not be used to help them out of this difficulty? The ring undoubtedly consisted of watery vapours and besides the advantage which it might furnish to the first inhabitants of the Earth, it had further this property of being able to be broken up on occasion, if need were, to punish the world which had made itself unworthy of such beauty, with a Deluge.

We thus are presented with the image of a circumterrestrial ring system, made, like its Saturnian cousin, out of individually orbiting satellites, and composed of water. In the *Theory of the Heavens*, he actually proposes bodies made of water – whether solid, liquid, or gas – in space, near the Earth. But he never considers that comets might be made of ice, although he comes tantalizingly close.

Kant argued that comets condense beyond the orbit of Saturn in a cloud of bodies with high eccentricities and all inclinations; that when they find themselves in the inner solar system approaching the orbit of the Earth, the Sun's heat warms them and vaporizes their surfaces; and that the tails are made of this vapour and driven back by some electrical influence of the Sun. Even apart from Kant's near deduction that comets are made of ice, this is a hard description to beat for 1755.

Kant's book was printed when he was thirty-one years old, and dedicated to Frederick the Great. But copies were never distributed to the Prussian Emperor or to anyone else: the publisher went bankrupt just as the book was coming off the presses. The dedication, craven in its submissiveness to secular authority, was typical of its day. It is reproduced in the box opposite. In fact, Kant was not an enthusiastic supporter of Frederick, and later in the century would express strong sympathies for the American and French anti-monarchal revolutions. He complained more than once that the state spent too much on war and too little on education.

He was also very cautious on religious matters. It bothered him that a natural explanation of the evolution of the solar system based only on Newtonian physics might be offensive to the prevailing faiths. He correctly predicted that adherents of established religions would argue as follows:

> If the structure of the world with all its order and beauty, is only an effect of matter left to its own universal laws of motion, and if the blind mechanics of the natural forces can evolve so glorious a product out of chaos, and can attain to such perfection of themselves, then the proof of the Divine Author which is drawn from the spectacle of the beauty of the universe wholly loses its force. Nature is thus sufficient for itself; the Divine government is unnecessary...

That is, the truth of how the universe works might be dangerous if it worked to undo sectarian teachings. The argument is with us still. In the light of this Kant says, 'I did not enter on the prosecution of this undertaking until I saw myself in security regarding the duties of religion.' If he could not have reconciled his scientific ideas with conventional religious doctrine, he says, he would have suppressed the former.

Despite such cautious attitudes, in 1788, for the first time in his life, Kant became embroiled in a controversy that was simultaneously political and religious: Frederick's successor, Friedrich Wilhelm II, began a campaign to uproot the pernicious teachings of the Enlightenment which had brought science and rationalism to European culture. In 1794, Kant received a cabinet order deploring his 'misuse' of philosophy.

THE PHILOSOPHER AND THE KING

Dedication of Immanuel Kant's
Natural History and Theory of the Heavens
to Frederick the Great

To

THE MOST SERENE,

THE MOST POWERFUL KING AND LORD,

FREDERICK,

KING OF PRUSSIA, MARGRAVE OF BRANDENBURG, HIGH

CHANCELLOR AND ELECTOR OF THE HOLY ROMAN

EMPIRE, SOVEREIGN AND ARCH-DUKE OF

SILESIA, ETC.

MY MOST GRACIOUS KING AND LORD, MOST SERENE,
MOST MIGHTY KING, MOST GRACIOUS KING AND LORD!

The feeling of personal unworthiness and the splendour of the throne cannot make me so timid and fainthearted, but that the favour which the most gracious of Monarchs extends with equal magnanimity to all his subjects, inspires in me the hope that the boldness which I take upon me will not be regarded with ungracious eyes. With the most submissive respect I lay herewith at the feet of your Royal Majesty a very slight proof of that zeal with which the Academies of Your Highness are stimulated by the encouragement and protection of their enlightened Sovereign to emulate other nations in the sciences. How happy would I be, if the present Essay should succeed in obtaining the supreme approbation of our Monarch for the efforts with which the humblest and most respectful of his subjects has unceasingly striven to make himself in some measure serviceable to the good of his country. With the deepest devotion till death,

I am,

YOUR ROYAL MAJESTY'S

Most humble Servant,

THE AUTHOR.

KÖNIGSBERG,
14th March, 1755

Who would know more about the uses of philosophy, the Prussian Emperor or Immanuel Kant? His teachings were found to be insufficiently deferential to the prevailing theological wisdom, and he was warned explicitly: 'If you continue to oppose this order, you may certainly expect unpleasant consequences to yourself.' Kant soon submitted and attempted to justify his submission: 'Recantation and denial of one's inner convictions is base, but silence in a case like the present is a subject's duty. And if all that one says must be true, it does not follow that it is one's duty to tell publicly everything which is true.' On this issue, he was not a giant.

Kant's biographer, writing in 1899, makes a character assessment that is of some interest in light of the tragic history of Germany in the first half of the twentieth century:

> Perhaps we may say that there is an inner relationship between Kant's ethics and the Prussian nature. The conception of life as service, a disposition to order everything according to rule, a certain disbelief in human nature, and a kind of lack of the natural fullness of life, are traits common to both. It is a highly estimable type of human character which meets us here, but not a lovable one. It has something cold and severe about it that might well degenerate into external performance of duty, and hard doctrinaire morality.

Kant's celebrated philosophical endeavours began in an attempt to work out the general implications of the Newtonian vision of the world. Much of Kant's philosophy represented a running battle with the widely influential philosophy of Gottfried Wilhelm von Leibniz, a stilted, formalistic, and anthropocentric world-view with delusions of completeness. In his *Critique of Pure Reason*, Kant announced that he had worked a Copernican revolution in philosophy. Copernicus had shown that the apparent motion of the Sun, the Moon, and the stars was in fact due to the motion of the observer. Some issues – including immortality, freedom, and God – Leibniz had asserted were knowable; but Kant argued that by their very natures, they could not be fully experienced, and were in fact not known; Leibniz had presented the illusion and not the reality of knowing. This was heady stuff, and the authorities of course considered it subversive.

Kant's role in the history of philosophy is generally considered magisterial, but we wonder – we recognize that this is a heretical view – whether he might have been more effective in the long run had he continued the remarkable scientific work of his youth and left the metaphysics to others. Inscribed on Kant's tomb are his words:

> The starry heavens above me and the moral law within me fill the mind with ever new and increasing admiration and awe.

Edmond Halley had predicted that the comet of 1682 would return at the end of 1758. It is easy to see why his prophecy caused little excitement at the time – 1758 was then more than half a century in the future. When Halley died in 1742, his obituaries made no mention that he had predicted the return of a comet. Instead, his voyages of exploration and the diving bell he invented were, however, given great emphasis.

But by 1757, there were those who had become obsessed with the idea that the gravitational physics of Newton could actually be used to foretell the future. Among them was Alexis Clairaut, an eminent French mathematician, who had published his first paper at the age of thirteen. He made a last minute decision to try to improve on Halley's tables on the orbit of the comet of 1682, and the predicted time of its return. It was imperative, of course, that the revised prediction appear before the comet did, 'so that no one might doubt the agreement between the observation and the calculations'. But the comet was fast approaching and the task was enormous, requiring meticulous calculation of the gravitational inter-actions of Jupiter, Saturn, the Earth and the comet over a period of 150 years. Clairaut claimed that he engaged the astronomer Joseph Jerome de Lalande to help him. To hear Lalande tell it, it was the other way around. But Clairaut made no public mention of the third member of the team, without whom – as Lalande was later to admit – they would never have dared to try to beat the comet; it was she who deserved most of the credit.

We can only imagine the forebearance that Madame Nicole-Reine Etable de la Briere Lepaute needed to get through her remarkable life. It was an age when upper class women were valued for their appearance, their ability to oversee the running of the household, and their capacity for lively small talk. Madame Lepaute fulfilled these ideals, but she was also a first-rate mathematician. In this regard, she

4.9 *Chart of the apparent path across the sky of the celebrated comet of 1759. (From* Histoire de l'Académie Royale des Sciences, 1760, Paris, Impr. Royale, 1766.)

4.10 *Nicole Lepaute. (Portrait by Voiriot. Courtesy, Michel-Henri Lepaute.)*

4.11 *Extra! Extra! Halley's Comet Returns on Schedule! This is the substance, but far from a literal translation, of the front page of the* Hamburgisches Magazin *of late January 1759.*

posed a problem for her colleagues that is made clear by Lalande's tribute to her in his *Astronomical Bibliography.* Writing soon after her death in 1788, Lalande goes to great lengths both to exalt and to belittle her. Yes, she was vital to their work on the comet, but she was not pretty enough. Yes, her tables of parallactic angles, and her accurate prediction for the whole of Europe of the annular eclipse of 1764, were important, but no, she herself was significant mainly in terms of her male relatives. He *will* give her this much: 'Her calculations never got in the way of her household affairs; the ledgers were next to the astronomical tables.' Clairaut suppressed knowledge of Mme Lepaute's contribution, Lalande wrote, 'in order to accommodate a woman jealous of Mme Lepaute's merit, and who had pretentions but no knowledge whatsoever. She was able to have this injustice committed by a judicious but weak scientist whom she had subjugated.' Scientific biographies were racier then.

But Lalande's devotion to her is clear:

> Mme Lepaute's was the only woman in France who acquired true insights into astronomy ... She was so dear to me, that the day I walked in her funeral procession was the saddest I have spent since I learned of the death of my father ... The times I spent near her and in the heart of her family are those I am most fond of, the memory of which, mixed with bitterness and pain, spreads some comfort over the last years of my life ... Her portrait, which I still have before my eyes, is my consolation.

Lepaute must have had her hands full in 1757. With Clairaut and Lalande, she worked day and night, often through meals, for six months, in a desperate race with the comet. The enterprise was so taxing, Lalande later wrote, 'that following this forced labour, I contracted an illness which would change my temperament for the rest of my life'. Eventually, they discovered that the comet would be detained by Saturn's gravity by 100 days. Jupiter meant a delay of at least 518 days. In the course of their calculations, they found that Halley had made a set of compensating errors which cancelled each other out, and concluded that Halley's estimate of the time of the return was essentially correct.

In November of 1758, they predicted that the comet would achieve perihelion passage on 13 April 1759 and might be visible some months before. On Christmas night 1758, a German farmer – one Johann Palitzsch – became the first to know that the long-dead Edmond Halley had successfully employed Newton's Laws to foretell the future. The comet was punctual, and it came from just the sector of the sky that Halley had foretold. Palitzsch, an avid amateur astronomer, one of many to make a contribution to cometary astronomy,

4.12 *Halley's Comet in 1759, as painted by Samuel Scott.*

rushed to tell the world. Halley's prodigal comet had returned. It reached perihelion on 13 March 1759, within a month of the Clairaut-Lalande-Lepaute prediction. Science had succeeded where generations of mystics had failed. Newtonian prophecy had been fulfilled.

Many soon recognized what Halley and his French successors had accomplished. They had established a programme, a goal, an ideal for the future of all of science: 'The regularity which astronomy shows us in the movements of the comets,' Laplace concluded, 'doubtless exists also in all phenomena.'

Chapter 5

ROGUE COMETS

Thou too, O Comet, beautiful and fierce,
Who drew the heart of this frail Universe
Towards thine own; till, wrecked in that
    convulsion,
Alternating attraction and repulsion,
Thine went astray, and that was rent in
    twain;
Oh, float into our azure heaven again!

           Percy Shelley,
           Epipsychidion
           (1821).

The triumphant return of Halley's Comet in 1758 powerfully supported, in the minds of people all over the world, the Newtonian view that we live in a clockwork universe. In the predictable motion of the planets, and in the periodic apparitions of Comet Halley (and, later, its brethren), many people saw the hand of God. Seeking new comets and making preliminary determinations of their orbits became a fashionable avocation. In the time of the American and French revolutions, optimistically proclaimed The Age of Reason, the regular motions of the comets represented a continuing reminder of the gradual emergence of the human species from rank superstition, while at the same time the majesty and elegance of a Divine Purpose was considered evident in every cometary orbit.

As a larger sample of comets was examined, however, some odd idiosyncrasies, and a disquieting departure from Newtonian regularity, were uncovered. A class of short-period comets was found. They circuit the Sun once every few years, entirely among the planets in the inner solar system. For example, Comet Encke, discovered in 1786, comes closer to the Sun than does the innermost planet, Mercury; once during each orbit. In 1819, J. F. Encke was studying the repeated returns of the comet that now bears his name. Since the period is only 3.3 years, he had a number of orbits to contemplate. Encke found, to his considerable surprise, that in every perihelion passage the comet was arriving a couple of hours early – even after perturbations by Jupiter and the other planets were properly taken into account. Here was a substantial mystery, which Encke was destined never to solve.

5.1 *(Opposite) A cometary nucleus fragments as it passes by the planet Jupiter, seen here nearly pole-on. Each of the large fragments will develop its own tail as this swarm of irregular pieces of ice approaches the inner solar system. (Painting by Michael Carroll.)*

5.2 *Drawing of the coma of Encke's Comet, observed on 30 November 1828. (Courtesy, R.A. Lyttleton, from his book* The Comets and Their Origin, *Cambridge University Press, 1953.)*

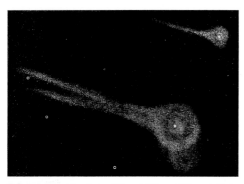

5.3 *A telescopic view of the splitting of Comet Biela. (Drawing by Struve from Amédée Guillemin,* The Heavens, *Paris, 1868.)*

5.4 *Drawing of the appearance, seen through the telescope, of the comet discovered by Biela and Gambert during its apparition of 1846. The comet had split in two since its previous apparition·in 1832. (From Camille Flammarion,* Astronomie Populaire, *1880.)*

It put the new astronomy in an awkward position: comets had been touted as the proof of a precise and universal law of gravitation, but at least one comet chose not to play along. Even Newton could not make all the comets run on time. The phrase 'defying the law of gravity' goes back to these times. Most scientists believed that Newtonian gravitation was valid, but that here some other, additional force was at play. But what?*

On 27 February 1826, a major in the Austrian army named Wilhelm von Biela found himself in South Africa looking up at the sky, whereupon he discovered what seemed to be a new comet. Ten days later, it was independently discovered in Marseilles by a French astronomer named Gambert, who promptly calculated the orbit and deduced a period of a little under seven years. An unpleasant and petty controversy arose between Biela and Gambert about the priority of the discovery. Should it be called Comet Biela, after the first person to see it on its most recent apparition, or Comet Gambert, after the first person to calculate its orbit? Meanwhile, Gambert made the ominous prediction that *his* comet would strike the Earth on its next return, around 29 October, 1832. Although the comet appeared on schedule, it did not strike the Earth. We can imagine some anxious moments (and for more reasons than one) around the Marseilles observatory in the fall of 1832.

On its next predicted perihelion passage, in 1839, the comet's position in the sky at closest approach to Earth was very near the Sun, and in the glare no comet was seen. But in 1846, the apparition was more favourable, and, as astronomers peered through their telescopes in amazement, it was discovered that now there was not one but two comets on almost identical trajectories, each with its own tail. During the next weeks the relative brightness of the two comets varied, first one and then the other shining more brilliantly. For a while there was even a common coma enveloping both. The finding was so bizarre that the first astronomer to note this twinning dismissed it as some internal reflection in his telescope. How a comet could reproduce itself was a mystery worth pondering, although it did at least provide a Solomonic resolution to the battle over priority between Biela and Gambert. In the 1852 apparition, two comets were still seen, but now some two million kilometres apart, although still travelling in approximately the same orbit. The two comets were never seen again. Comets that arrive early or late, comets

*Among the erroneous possibilities offered were friction by an enormous quantity of otherwise undetected interplanetary dust; the resistance of a postulated 'luminiferous aether', in which light waves were once thought to propagate; and a small departure of gravity from the inverse square law.

that split or reproduce themselves, comets that disappear – all of this undermines the notion that comets are wholly and exclusively subservient to the Newtonian clockwork.

Although Comets Biela/Gambert never returned, they nevertheless provided a further astonishment for the perplexed astronomers of Earth. Only a few decades after the comets were lost forever, there began on Earth the stunning November meteor shower called the Andromedids – thousands of brilliant 'falling stars' illuminating an autumn night. When the orbits of the Andromedid meteors were traced back, it was discovered that they had precisely the trajectory of Comets Biela/Gambert. Somehow, the comets had disintegrated, leaving in their place only a multitude of fine debris which entered the Earth's atmosphere when the orbits of the comet and our planet crossed. Since then, most of the prominent meteor showers have been connected with cometary orbits. Meteors – universally known as falling or shooting stars – visibly fall, dissipating in a few moments. Comets do not streak across the sky; and the brightest of them can be seen with the naked eye for months. Despite these differences, it was beginning to seem that meteors and comets were related. The idea took hold that comets were, in fact, self-gravitating swarms of fine particles: when they were all collected together, moving as a swarm, they were seen as comets; when they individually entered the Earth's atmosphere, they were seen as meteors. A model of the cometary nucleus as a kind of orbiting gravel bank took form.

In 1744, de Chesaux's Comet made a spectacular apparition. The tail was divided into six equal 'rays', and for a while Europeans could see the tail of the comet above the horizon, while the head (and the Sun) lay below. It was commemorated in lithographs, scientific diagrams, and even on the coinage. A general expectation was being established that every few decades a bright comet approaches the Earth.

The first short-period comet on record is Lexell's, which came very close to the Earth in the year 1770. Lexell calculated the orbit and derived a period of only a little more than five years. The comet passed so close to the Earth that the transient gravitational grasp of our planet caused a decrease in its period by almost three days – although the period of the much more massive Earth did not change by so much as a second in a year. In 1776 it was not seen, but this was attributed to its greater distance from the Earth than during the previous apparition. Wait until 1781, the astronomers assured the public. But 1781 came and went without the comet. In this case, no meteor stream was subsequently found in the orbit of the comet. What had happened?

The question was answered by Laplace. In effect, he

5.5 *Obverse face of a German medal, probably struck at Breslau in 1744, depicting de Cheseaux' Comet with its six tails. The comet is properly shown as brighter than the brightest stars; indeed, in March 1744, it could be seen in full daylight. (The American Numismatic Society, New York.)*

5.6 *Reverse of the German medal commemorating de Cheseaux' Comet. The inscription, from Romans 11, 34, reads: 'Who hath known the mind of the Lord?'*

5.7 *The tail of de Cheseaux' Comet of 1744 seen above the horizon, while the head is below the horizon. (From Amédée Guillemin's* The Heavens, *Paris, 1868.)*

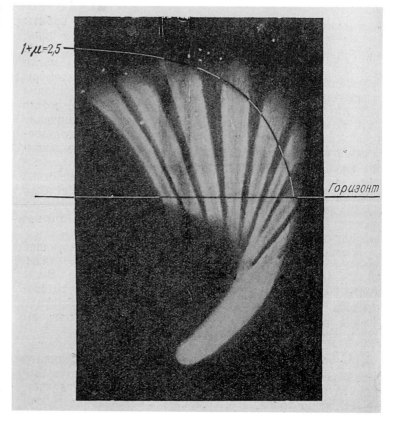

5.8 *A drawing of de Cheseaux' Comet of 1744, with the horizon indicated by the horizontal line. (From S.V. Orlov,* On the Nature of Comets, *Soviet Academy of Sciences, Moscow, 1960.)*

constructed a giant mathematical orrery*, and ran the motion of Lexell's Comet (and the planets) backwards and forwards in time from its solitary apparition in 1770. The calculations, especially in the era before computers, were not simple. Laplace learned that the comet did not reappear in 1781 because in 1779 it had passed extremely close to Jupiter, possibly threading its way among the giant planet's four large moons. Earlier, Halley had found that, passing much further from Jupiter, the orbital properties of his comet would be changed a little by the tug of Jupiter's gravity. But here, Laplace found that Lexell's Comet had passed so close to Jupiter that its orbit must have changed dramatically. The comet had been thrown into some alternative trajectory, coming nowhere near the Earth – indeed, perhaps being ejected from the solar system altogether. In its close approaches both to the Earth and to Jupiter, Comet Lexell would have provided a magnificent view to a hypothetical observer on board. Laplace could not help thinking that every now and then a comet might strike a planet, and this set him to wondering what the consequences would be on Earth if a comet ploughed into it. (See Chapter 15.)

By Laplace's time, in the beginning of the nineteenth century, there were three categories of comets known: the

5.9 *Inclinations of cometary orbits. The ecliptic, or zodiacal, plane, shown in dark blue, contains the nearly circular orbits of the planets (orbits of the Earth and Jupiter are shown). The short-period cometary orbits (shown in red) tend to lie in the zodiacal plane. The long-period cometary orbits (blue, not circles) tend to be randomly distributed. Some short-period orbits are inclined at large angles to the ecliptic plane and some long-period orbits happen, by chance, to lie in the ecliptic plane. (Diagram by Jon Lomberg/BPS.)*

*A mechanical model of the solar system with the period and/or distance of the moons and planets rendered to scale.

REPUBLIQUE FRANÇAISE

30F   +9F
POSTES

LAPLACE
1749-1827

LEMAGNY   COTTET

5.10 *Pierre Simon, Marquis de Laplace, French mathematician, physicist, and astronomer, who played a major role in understanding the nature of cometary orbits. Laplace's contributions to science were fundamental and varied; as a sideline, in 1780 with Lavoisier, he showed that respiration is a form of combustion.*

short-period comets (like Lexell, Encke and, later, Biela/Gambert) had periods of a few years and lived entirely in the inner solar system. The long-period comets (like Halley's or the Great Comet of 1680), with periods measured in decades or centuries, had orbits that took them out beyond the farthest known planet. But the bulk of the comets were 'new' – with orbital periods so long that they could not be determined from the available data. Were these three separate kingdoms of comets, made of different building materials and having different origins? Or were they related, members of one kingdom evolving into another? All over the Western world there was revolution in the air, and the idea that absolute monarchies could be quickly transmogrified into something like real democracies made many people wonder what additional changes, hitherto deemed unthinkable, might be possible in other realms of nature. It was a mark of Laplace's genius how successfully he thought in evolutionary terms when most of his contemporaries still considered the main features of the universe as – necessarily, as a mark of God's handiwork – unchanging through the eons.

Laplace proposed that Jupiter's gravity was a kind of net, capturing new and long-period comets that wandered near and transforming them into short-period comets, henceforth to reside in the inner solar system. A modern rendition of the argument goes something like this: short-period comets tend to have their aphelia, the far points from the Sun in their periodic orbits, near to the orbit of Jupiter. An even higher percentage have one node of their orbits – the point where the plane of the cometary orbit intersects the plane in which the planets move around the Sun – close to Jupiter (see Figure 5.11). In contrast, the orbits of the new and long-period comets show no connection with Jupiter. Also, some short-period comets – those restricted entirely to the realm of the terrestrial planets, for example – know nothing about Jupiter. But there are so many short-period comets with orbital characteristics tied to Jupiter that they are called the Jupiter family of comets.

What does Jupiter have to do with the family of comets associated with it? Some early astronomers thought this meant that Jupiter was the *source* of the comets, that they were somehow disgorged from the inside of the largest planet. The most generous thing we can say about this view is that it has not stood the test of time. What other possibilities are there?

The comet is barrelling in towards the Sun, approaching the orbit of Jupiter – as, it may be, it has done a dozen times before with Jupiter on the other side of the Sun and nothing untoward transpiring. But this time, by chance, as it crosses the orbit of Jupiter, Jupiter happens to be nearby. It is the most massive of the planets, and the comet is, by comparison, a

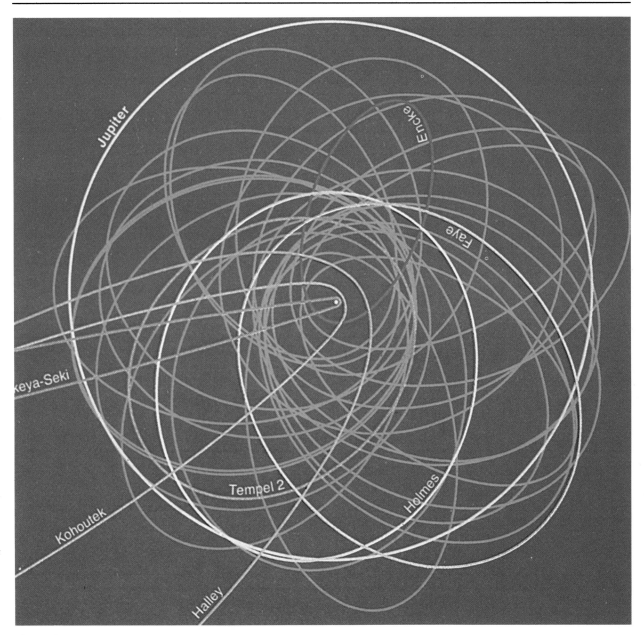

5.11 *A few of the known cometary orbits shown in relation to the orbit of Jupiter (large white circle). Among the comets shown are Encke's (in red), and Halley's. Note how frequently the aphelion of the orbit of a short-period comet lies near to the orbit of Jupiter. (Courtesy, NASA.)*

little thing, a puff of gas enveloping a speck of matter. Jupiter's gravity attracts the comet – not enough to draw it into Jupiter (the comet is travelling at tens of kilometres a second), but enough to deflect it towards Jupiter, and thus to change its orbit. The comet is still travelling in an ellipse about the Sun, but the gravitational encounter with Jupiter has dramatically changed the orbit (see Figure 5.12).

A similar manoeuvre was successfully carried out by four spacecraft from Earth – Pioneers 10 and 11, and Voyagers 1 and 2 – which in the 1970s were sent to Jupiter with so exquisitely crafted a trajectory that the gravitational acceleration of Jupiter swung each of them like a slingshot off toward a precisely

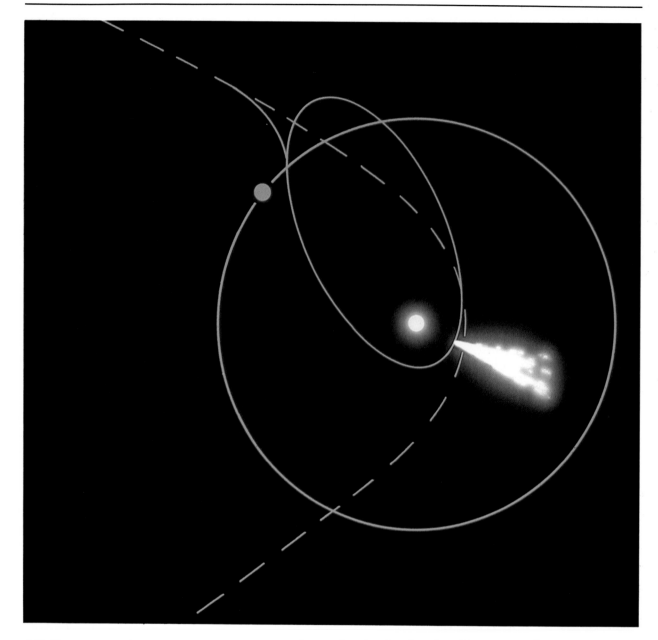

5.12 *The conversion of a long-period comet into a short-period comet through passage by the planet Jupiter. Jupiter is the orange dot shown in circular orbit around the Sun. The comet approaches the Sun on the dashed line from top left. Were it not for Jupiter, it would continue around the Sun and out into the distant reaches of the solar system, along the orbit extending bottom left. But Jupiter's attraction perturbs it into the solid elliptical orbit, as shown. The comet's tail is fully developed as it achieves perihelion passage. (Diagram by Jon Lomberg/BPS.)*

pre-planned point in the sky. Voyager 2, for example, was swung by Jupiter in such a way that it was propelled towards a close encounter with Saturn two years later. (The close encounter of Saturn was also designed to swing it by Uranus in 1986, and the Uranus encounter to swing it by Neptune in 1989.) Now imagine the Pioneer or Voyager orbits run backwards in time. The spacecraft approaches Jupiter from the outer solar system; it races around Jupiter, and is carried on its new orbit to approach the Earth.

This kind of gravitational billiards explains the Jupiter family of comets. The inner solar system is sprayed with many comets, a few of which by accident come close to Jupiter. Some are immediately ejected by the encounter out of the solar system, some run into Jupiter or one of its moons, but many others have their orbits converted so that they become short-period comets of low inclination, with aphelion and node near Jupiter's orbit. Most short-period comets may have achieved their orbits by multiple gravitational encounters with Jupiter, or even by multiple encounters with planets exterior to Jupiter, and then with Jupiter itself.

A comet arriving from the depths of space in a direct orbit – one that goes around the Sun in the same direction as the planets (clockwise as seen from above the North Pole) – is more likely to be injected into a short-period orbit by an encounter with Jupiter. A comet coming in with a retrograde orbit – going around the Sun in the opposite direction – is more likely to be ejected from the solar system in the encounter. This is why the motion of short-period comets is, without exception, direct. The comets being supplied from the depths of space are as likely to be on retrograde orbits as direct, but only the comets in direct orbits tend to be captured. Jupiter's gravity makes distinctions.

It is estimated that there are about 2,000 long-period comets bigger than a kilometre that cross Jupiter's orbit every year. The number of comparably-sized short-period comets in the Jupiter family is about 1,400. When we talk of the orbital evolution of the comets we see, we are talking about thousands of worlds.

Laplace showed that new and long-period comets could be converted into short-period comets by a gravitational machinery working before our very eyes. His work also seemed to demonstrate that interstellar comets, visitors from beyond the solar system, could be converted into short-period comets about the Sun too.

A few comets have been seen in their passage through the inner solar system on very slightly hyperbolic trajectories – untied from the solar system, bound for the stars. It is natural to think of them as interstellar nomads, perhaps comets from some other star system, long wandering through interstellar space and by lucky chance just passing through on our watch. Indeed, Laplace believed that both the short- and the long-period comets, as well as many 'new' comets, were captured by the Sun from a population of free interstellar comets surrounding the Sun. If so, then some short-period comets might have evolved through an orbital cascade, a succession of changing trajectories determined by successive planetary encounters. The implication, which was hardly lost on Laplace, was that comets are essentially denizens of the interstellar cold and dark, a fact fundamental to the modern understanding of comets.

5.13 *Halley's Comet as a reckless driver streaking past the planets. Cartoon by Hermann Vogel. (From* Fliegende Blätter, *May 1910.)*

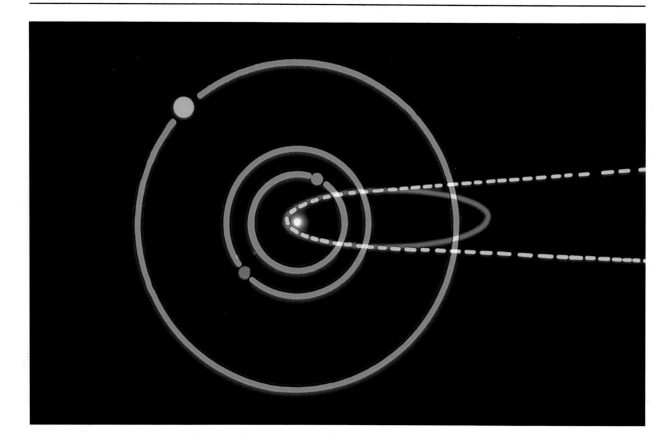

5.14 *If we are able to track a comet only when it is close to the Sun, we may be unable to determine its orbit. Here, the concentric circles show the orbits of the Earth, Mars, and Jupiter. A comet on a parabolic or hyperbolic orbit, coming from interstellar space (dashed yellow line), may not be distinguishable in the inner solar system from a comet on an elliptical orbit (red closed curve). (Diagram by Jon Lomberg/BPS.)*

Nevertheless, despite their unconstrained trajectories, these hyperbolic comets turn out not to come from interstellar space, at least none observed so far. Using the same sort of mathematical orrery that Laplace pioneered, we can track the orbits of the apparently hyperbolic comets back in time. Remarkably, every such hyperbolic comet is only slightly hyperbolic; if it were moving only a little more slowly it would be gravitationally bound to the Sun. When we track the orbits of such hyperbolic comets, we find, in their recent past, a close approach to one of the major planets. All of them seem to be comets that have been on long elliptical orbits about the Sun, and then slingshot out of the solar system by a close gravitational encounter with Jupiter or another of the giant planets. Not a single interstellar comet has ever been observed.

By the middle of the nineteenth century, the study of comets was an established part of professional astronomy. Their fundamental motions were well-understood. Some new and long-period comets might eventually evolve into short-period comets, and in the tumult of gravitational encounters, comets might collide with the planets or the Sun or be ejected from the solar system. There was something chaotic, slightly unsettling about the motions of the comets. Some fell to pieces unexpectedly, at least one delivered itself of small, non-Newtonian feints and lurches, interrupting the stately circumsolar procession. But by and large, comets were considered well-understood. By now it was possible to illustrate public lectures with photographs of comets and some of these stirred the multitudes. For a glimpse of what a professional astronomer, expert on comets and with a flair for lucid explication, was saying to the general public on the subject, the following is an excerpt from a talk delivered in 1882 by William Huggins at one of the public lectures (Friday Evening Discourses), held then and now for the general public at The Royal Institution, London:

5.15 *The Great Comet of 1882, in what may be the earliest successful photograph ever obtained of a comet. (Photographed by David Gill in South Africa.)*

> With the aid of a telescope, in the heads of most comets a minute bright point may be found. The apparently insignificant speck is truly the heart and kernel of the whole thing – potentially it is the comet. It is this small part alone which conforms rigorously to the laws of gravitation … If we could see a great comet during its distant wanderings when it has put off the gala trappings of perihelion, it would be a very sober object, and consist of little more than nucleus alone … Under the Sun's influence, luminous jets issue from the matter of the nucleus on the side exposed to the Sun's heat. These are almost immediately arrested in their motion Sunwards, and form a luminous cap; the matter of this cap then appears to stream out into the tail, as if by a violent wind setting against it. Now, one hypothesis supposes these appearances to correspond to the real state of things in the comet, and that there exists a repulsive force of some kind acting between the Sun and the gaseous matter, after it has been emitted by the nucleus … Great electrical disturbances are set up by the Sun's action in connection with the vaporization of some of the matter of the nucleus, and … the tail is matter carried away, possibly in connection with electrical discharges, in consequence of the repulsive influence of the Sun …
>
> A comet would, of course, suffer a large waste of material at each return to perihelion, as the nucleus would be unable to gather up again to itself the scattered matter of the tail: and this view is in accordance with the fact that no comet of short period has a tail of any considerable magnitude.

Virtually every one of these remarks of Huggins is in reasonable accord with the modern understanding of comets. Some were far ahead of the knowledge of his time. The subject was considered respectable and mature. Nevertheless, the central fact about the comets – their composition, and the nature of the spectacular variations in the appearance of comets – was at best only dimly glimpsed. In reading the literature of the time, we are struck by how rarely it was even acknowledged that these were important matters awaiting future discovery.

Chapter 6

ICE

The frightful ice that covers the whole face of the land.

Hans Eggede,
A Description of Greenland
(1745).

One of the central questions about comets, and surely the key to many of its mysteries, is their composition. What is a comet made of? Are they all made of the same stuff? In the sixteenth and seventeenth centuries, it was still customary to think of comets as Aristotle had – as gasses, vapours, 'exhalations' from the Earth, and perhaps from the Sun and the planets as well. Newton, clearsighted as usual, thought otherwise. He noticed that the comet of 1680 came very close to the Sun; at perihelion, 0.006 of an Astronomical Unit, a little under a million kilometres. This should, he estimated, have heated the comet to the temperature of red-hot iron, from which he deduced that the comet could not be composed only of vapours and exhalations – because then its substance would have been rapidly dissipated during perihelion passage. Instead, he concluded that the 'bodies of comets are solid, compact, fixed, and durable, like the bodies of the planets'. Since the tail of this comet was 'much more splendid' just after perihelion than before, Newton concluded that the heat of the Sun produced the tail: 'The tail is nothing else but a very fine vapour, which the head or nucleus of the comet emits by its heat.'

Fair enough. But what is the nucleus made of? Just what is this 'very fine vapour' that constitutes the tail? This is the problem with which Kant and many others had tentatively grappled. When a comet comes as close to the Sun as the Great Comet of 1680 did, then virtually any common material should start to vaporize. But many comets begin to develop comas and tails when they are between the orbits of Mars and Jupiter. Heated only by sunlight, the temperatures of these comets, as they begin to pour vapour out into space, are something like a hundred degrees below zero on the Centigrade scale. Materials like iron that do not vaporize until they reach a high temperature are called involatile or refractory. Materials like ice that turn into gas after relatively modest heating are called volatile. The comets, therefore, must be composed of something quite volatile. But what?

6.1 *(Opposite) Two exquisite manifestations of ice: a comet over Antarctica. (Painting by Kim Poor.)*

6.2 *Comet Pons-Winnecke, as sketched by Baldet. (Courtesy, R.A. Lyttleton, from his book* The Comets and Their Origin, *Cambridge University Press, 1953.)*

The propensity of some comets to split suggests a nucleus not very strongly compacted. The forces that hold it together must be rather weak. As we have seen, every now and then as the comet approaches the Sun there is a very small darting motion, entirely at variance with the languid Newtonian sweep that the comets usually exhibit as they fall in towards the Sun. These erratic and unpredictable non-gravitational motions of the comets recall Kepler's image of comets as fishes darting through the cosmic ocean. Encke described his comet as deviating 'wildly'* from the motion predicted by Newtonian gravitation, and attributed the anomalous and unpredictable movements to some resisting gas in interplanetary space, retarding its motion. But the accelerations are much too abrupt, and we now know there is not nearly enough material between the planets to have any detectable effect on the motion of comets. Some very different explanation is needed.

Until quite recently, the prevailing cometary fashions were dominated by the known association between comets and meteor streams. As late as 1945 the leading American college textbook in astronomy accepted without question the idea that comets are 'loose swarms of separate particles moving on parallel orbits through interplanetary space'. Some scientists believed this swarm of small meteors, imagined to make up the nucleus of the comet, to be gravitationally bound; others thought there was not enough mass in the nucleus to hold it together, and that instead there were merely an enormous number of small particles travelling in very closely the same orbit through space. There was a certain tendency for advocates of this flying gravel bank model to favour pointillist renditions of the comet head; some representative examples are shown in the drawings on this page.

6.3 *A pointillist rendition of the appearance of Halley's Comet in 1835. (Courtesy, R.A. Lyttleton, from his book* The Comets and Their Origin, *Cambridge University Press, 1953.)*

The gravel or sand bank hypothesis explained very nicely why an ageing comet might one day leave only a cloud of fine particles in its wake. The spectra of meteors, as they burn up in the Earth's atmosphere, show the presence of such materials as iron, magnesium, aluminium, and silicon, typical constituents of rocks on Earth. If meteors are made of rocky stuff, and comets in turn are made mainly of meteors, it followed that comets were rocks and stones. But what about the coma and tail? It was widely postulated that the sand particles were coated by some more volatile solid that evaporated as the swarm approached the Sun, or perhaps gasses were baked out of the stones as they were heated. But it

---

*An exaggeration. The non-gravitational forces change the period of Encke's comet by around one day each orbit. The orbital period is 1,200 days. So even to measure the effect requires an error less than 1/2000, or about 0.05%, not far short of perfect.

was hard to see how much of this material – whatever it was – could be left after a single passage by the Sun, if in the first place it was only a thin coating around a grain of sand, or the gas absorbed in rock. And there were other difficulties – the jet fountains, for example, which have no ready explanation from a loosely bound swarm of gravel.

For Comet Encke, at least, the issue is at last definitively resolved: when probed by radar from a large radio telescope on the surface of the Earth, it shows a single solid nucleus, not a swarm of particles. The size of the nucleus detected by radar – roughly a kilometre in radius – is consistent with other estimates. The nuclei of several other comets have also been detected by radar observations – that, accordingly, appear to sound the death knell for the orbiting gravel bank hypothesis of the comet nucleus. But before 1950, even the idea of a compact cometary nucleus was disreputable, and the nature of the cometary volatiles only vaguely glimpsed.

6.4 *Fred L. Whipple, leading proponent of the dirty ice model of the cometary nucleus.*

Fred Whipple describes himself as an Iowa farm boy turned astronomer. He served as Chairman of the Department of Astronomy at Harvard University and, for many years, as Director of the Smithsonian Institution's Astrophysical Observatory in Cambridge, Massachusetts. He had for years been thinking about small objects in the solar system, including the physics of meteors entering the Earth's atmosphere, and the nature of comets (of which he has discovered half a dozen). By the late 1940s, Whipple was convinced that large amounts of material poured out of comets near perihelion – much more than could be accounted for by ice coatings on grains of sand, or the driving out of small amounts of vapour that might be trapped inside the sand. It was also clear that interplanetary space was too good a vacuum for comets to be able to refurbish their supply of volatiles when their orbits took them far from the Sun. The problem was especially severe for comets like Encke, which repeatedly become very warm as they pass close to the Sun, and that seem to have made many perihelion passages.

As a convenient shorthand, Whipple called the refractory materials 'dust' and the volatile materials 'ice', and decided that the problem would be solved if there were a great deal more ice than the sand bank model of the cometary nucleus admitted. He describes the idea as 'obvious', although, remarkably, it had never been stated this baldly before. Such luminaries as Newton, Kant, and Laplace had all toyed with something similar, but it was Whipple who first stated the idea lucidly and coherently. He then showed that a number of other mysteries – including the splitting of comets, their dissipation into meteor showers, and the worrisome

ANTECEDENTS OF COMETARY ICE

Newton had obliquely implied that comets are made mostly of water (see Chapter 17), and Laplace mentioned casually that comets might be made of ice. They did not spell out their arguments, and these suggestions were mainly forgotten. But by the middle of the twentieth century, the idea was in the air again. For example, in the *Annales d'Astrophysique*, in 1948, there is an article by the Belgian astronomer Pol Swings, who says (rough translation) 'At large distances from the Sun, all the solids in a comet are at very low temperatures and all the "gas" that they contain, except for hydrogen and helium, must be found in the solid state.' A footnote, however, indicates that this insight earlier appeared in an article by the German astronomer, K. Würm, writing in the *Mittgang Hamburger Sternwacht Bergedorf*, in 1943. But in Würm's article, the comment is attributed to a remark made informally to Würm by the Czech/German chemist, Paul Harteck. In 1942 and '43 Harteck was busy trying to build an atomic bomb for the Nazis. His comments on cometary ices were perhaps a distraction from other, more pressing duties.

A summary describes briefly the observational success of the [icy conglomerate] comet model, both quantitatively and qualitatively. The surprising aspect of the model is its usefulness in spite of its vagueness...

Fred L. Whipple,
Present Status of the Icy Conglomerate Model
(Harvard/Smithsonian Center for Astrophysics, reprinted 1966, 1984).

non-gravitational forces acting on cometary motion – could all be explained if we revised our thinking and imagined the cometary nucleus as a ball of dirty ice, ice with mineral grains and perhaps other materials scattered throughout.

If comets are indeed made of dirty ice, then to understand comets we must understand something about ice. To begin, let's suppose that a comet is made of ordinary water ice. There are some ninety-two kinds of atoms occurring in nature, the most abundant of which are hydrogen, helium, oxygen, carbon, and nitrogen. These atoms combine with each other according to specific laws that are collectively called chemistry. Because there is more hydrogen than any other kind of atom in the universe, typical pieces of cold cosmic matter tend to be rich in hydrogen. Atoms such as oxygen, carbon, and nitrogen are often attached to as many hydrogen atoms as they can accommodate. Oxygen, for example, likes to combine with two hydrogen atoms, forming a molecule symbolized H_2O. The H, of course, stands for hydrogen, and O for oxygen, and the molecule in question, justly famous, is called water. A nitrogen atom likes to combine with three hydrogens, forming NH_3, also called ammonia; and carbon likes to combine with four hydrogen atoms, forming CH_4, a molecule also called methane. These abbreviations are a kind of shorthand picture of how the atoms are combined to form a molecule. There are many other simple combinations: CO, carbon monoxide; CO_2, carbon dioxide; HCN, hydrogen cyanide ... and an enormous variety of more complicated*

*The first of these is called acetic acid. The names of the other two are almost unpronounceable, and much longer than their formulas for diagrammatic representations of all known intersteller molecules, see Figures 8.12 and 13.

6.5 *Three molecules of water vapour, each containing an oxygen atom and two smaller hydrogen atoms. They are not linked together, but are freely moving as part of a gas. They might be molecules of steam released by a teapot. (Painting by Jon Lomberg.)*

molecules, such as $HCOOCH_3$, CH_3CCCN and $HC_{10}CH$.

Atoms that are a little rarer in the universe also make chemical combinations. If the silicon atom is symbolized by Si, then ordinary quartz sand is made up of SiO_2, silicon dioxide. The molecular oxygen in the air is O_2 (as distinguished from an oxygen atom by itself, O). And in this way the things of the natural world can be understood in terms of their constituent atoms. We humans are also aggregations of atoms, intricately and wonderfully assembled.

Now in a molecule like water, the atoms are not stuck onto each other at random positions or angles. An isolated water molecule always has the two small hydrogen atoms stuck on to the bigger oxygen atom at a precise angle, making the molecule look a little like a face with big ears, something like Mickey Mouse (see Figure 6.5). The chemical bonds that join atoms together do so according to precise and invariable rules which underlie the beauty and order of nature. These rules engage the cloud of electrons that surround every atom, but for the moment we treat atoms as little impenetrable balls. If you bring two of them near each other they will tend to combine according to certain definite rules, but if you try to push them too close together they will repel one another. A single isolated water molecule – exhaled from your mouth on a winter's day, for example – goes bobbing and tumbling off into the air happily colliding with the other molecules in front of it, and bouncing off them. It does not readily combine with those molecules; their constituent atoms are already preoccupied in passionate chemical embraces with other atoms.

Temperature is nothing more than a measure of the motion of molecules. When the temperature is high, the molecules

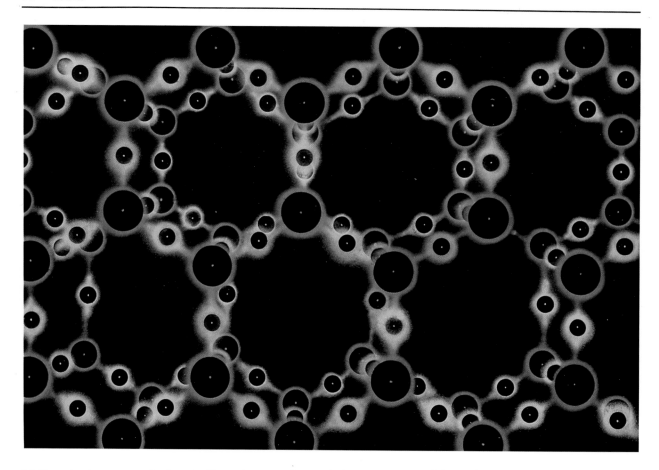

6.6 *The molecular structure of ordinary ice. The circles represent the electron clouds of individual atoms, with the tiny atomic nucleus inside. Large orange circles represent oxygen atoms, smaller yellow circles represent hydrogen atoms. The molecular forces between these atoms assemble them into this hexagonal crystal lattice. We see two of the many planes that would comprise a small piece of ice. (Painting by Jon Lomberg.)*

are in a riotous state, rushing about, tumbling, colliding, bouncing back – a frenzy of activity. As the temperature drops the molecules become more restrained, more decorous, almost sedate. At sufficiently low temperatures, water molecules connect – they are moving so slowly that the short-range molecular forces can engage, and adjacent water molecules join weakly to each other. When this happens – on a rainy day, for example – we say that the gas has condensed into liquid water. At lower temperatures still (below the freezing point), the molecules condense not in a helter-skelter order, but in an elegant repetitive pattern called a crystal lattice. This is the hidden structure of ice. In such a lattice every water molecule is in a specific place, joined to its neighbours as shown in the accompanying illustration.

In two dimensions, the pattern is hexagonal, like the tiles on a bathroom floor. Every hexagon is made of six oxygen atoms and the accompanying little hydrogen atoms. If you had a superior sort of microscope you would see such a structure along the surface and especially inside a surface of ice. You coud travel a million oxygen atoms to the left or the right and find exactly the same structure. In three dimensions the crystal lattice is a kind of hexagonal cage. Beyond the oxygen hexagon in front of us is another oxygen hexagon connected by the little hydrogen atoms. Beyond that is another and so on. You could rotate your head almost 120° and not see any discernible difference in the pattern. This hexagonal symmetry on the molecular level carries all the way up to the macroscopic level and is responsible for the exquisite six-sided symmetry of snowflakes (see Figure 6.6), a truth first glimpsed by the astronomer, Johannes Kepler.

6.7 *Side view of the atoms in a silicate crystal lattice. If you were to view this from the top, you would see a hexagonal structure. (Painting by Jon Lomberg.)*

Since the laws of nature are, as far we know, the same everywhere in the universe, this same hexagonal cage structure must apply to ice everywhere – on the surfaces of the moons of Uranus, in the particles of the rings of Saturn, in the tiny grains of interstellar space, and on the nuclei of the comets.

If we were to look closely at such a lattice, we would notice that the constituent atoms were not stationary, but instead were vibrating and throbbing in place. If we decreased the temperature, the throbbing would moderate; if we increased the temperature, the throbbing would become more violent. There is a characteristic strength of the chemical bonds that hold the ice crystal lattice together. At a certain temperature the constituent atoms are throbbing so violently that some of the bonds become broken and a small piece of crystal lattice – an isolated water molecule – detaches itself from its fellows and goes tumbling off. Statistically, this occurs even at low temperatures, on occasion. As the temperature increases, it occurs more often. At high enough temperatures, the throbbing motion becomes so violent that the upper layers of ice become disengaged and large numbers of individual molecules go gushing off the ice crystal. If this happens on a cometary nucleus, the water molecules then find themselves in nearby interplanetary space. The process is called, variously, evaporation, vaporization, volatilization, or sublimation. Fundamentally, it is a change from water in the solid state, called ice, directly to water in the gaseous state, called vapour, without experiencing an intermediate liquid state.

If we had a piece of ice in a closed container and heated it up, at the characteristic temperature the water molecules would start spewing off in earnest. But they could not escape to space; they would bounce off the walls of our chamber, and some of them would bounce back into the surface of the ice where they would tend to stick. This then would establish an equilibrium between molecules of water coming off the ice and molecules of water returning to it. Under these circumstances the rate at which the ice would be transformed into water vapour would be slow, and it is these circumstances with which we are familiar on the Earth. An ice cube in a covered jar disappears more slowly than one in the open air. But on a comet there is no air surrounding the ice, only a nearly perfect vacuum. So once the ice of the comet is sufficiently warmed, it begins to lose water molecules to space quickly, and forever.

Water ice isn't the only kind of ice there is. If you take ammonia gas or methane gas and sufficiently cool them they will also form a cold, white crystalline solid. The temperature

```
┌─────────────────────────────────────────────────────┐
│                                                     │
│            FREEZING POINTS OF ICES                  │
│                                                     │
│                                                     │
│        Water, H₂O                    0°C            │
│        Hydrogen Cyanide, HCN        −14°            │
│        Carbon Dioxide, CO₂          −57°            │
│        Ammonia, NH₃                 −78°            │
│        Formaldehyde, HCHO           −92°            │
│        Methane, CH₄                −182°            │
│        Carbon Monoxide, CO         −199°            │
│        Nitrogen, N₂                −210°            │
│                                                     │
│                                                     │
│    Temperatures are given in degrees Centigrade, or Celsius. │
│                                                     │
└─────────────────────────────────────────────────────┘
```

FREEZING POINTS OF ICES	
Water, H_2O	0°C
Hydrogen Cyanide, HCN	−14°
Carbon Dioxide, CO_2	−57°
Ammonia, NH_3	−78°
Formaldehyde, HCHO	−92°
Methane, CH_4	−182°
Carbon Monoxide, CO	−199°
Nitrogen, N_2	−210°

Temperatures are given in degrees Centigrade, or Celsius.

at which these gasses form ices is called the freezing point, a different temperature for each material.* The freezing points of some common gasses are shown in the adjacent table. The fact that water molecules crystallize at a comparatively high temperature is due to the strength of its chemical bonds; methane ice or ammonia ice, with their comparatively weak chemical bonds, fall to pieces at temperatures where water ice is quite stable and happy.

Now, imagine some mixture of methane, ammonia, water, and other gas molecules, but with water in excess, as the general cosmic abundances suggest should be common. Most of the universe is, of course, far from stars, and therefore well below the freezing point of water. On some tiny grain of dust out beyond Saturn, or perhaps between the stars, the temperatures are very low, so a molecule of water hitting the grain will stick. As the grain passes over millions of years through a thin gas, it will grow, and the hexagonal crystal lattice will, bit by bit, extend itself in all directions. As it does, other sorts of molecules may become trapped in the hexagonal cage – a molecule of methane, perhaps, or ammonia, or something else (see Figure 6.9). This kind of water ice structure, in which foreign molecules are trapped inside the cage, is called a clathrate. The trapped molecule is not chemically combined with the ice, but is merely imprisoned physically. You can see that there is room for only about one big atom in the cage, and so such clathrates tend to have around six water molecules for every methane molecule. In the vaporization of a water clathrate, the imprisoned

*Depending on the local atmospheric pressure, gasses can freeze solid without passing through an intermediate liquid stage; this is true for water vapour on Mars or an interstellar grain, but not at the surface of the Earth where the atmospheric pressure is large.

6.9 *A molecule of methane sits trapped in one of the hexagonal cages of an ice crystal lattice. The carbon atom is shown with white outlines, attached to four smaller hydrogen atoms, comprising methane, CH₄. Ices with such trapped molecules inside them are called clathrates. (Painting by Jon Lomberg.)*

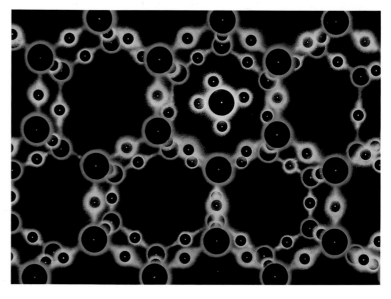

molecules are released to space at the same time that the ice crystal lattice peels off.

If this grain is growing over millions of years, it may be composed largely of water ice, because there is so much more water than anything else that will condense onto the grain. But other materials should be trapped as clathrates, and other ices should form on its surface – patches of CH_4 ice or NH_3 ice or CO_2 ice, as well as much more complex molecules. If the grains continue to grow and collide with one another, building structures of increasing size, eventually something like a small cometary nucleus will evolve. It will not be purely ice. There are many other kinds of material available, some of which we will mention later. But for now, let us imagine the cometary nucleus composed only of ices.

Picture the cometary nucleus plummeting into the inner solar system. The temperature at its surface slowly warms. Because ice is not a good conductor of heat, the interior of the cometary nucleus remains for a long time at the temperature of the interstellar cold. But the outside becomes steadily hotter. Eventually, it gets so hot that the chemical bonds holding the ice together begin to break, and the outer layers of ice go spewing off the comet into space. Different ices experience this violent vaporization at different temperatures, and therefore at different distances from the Sun. As the comet crosses the orbit of Neptune, a patch of pure methane ice is feebly warmed by the approaching Sun; as chemical bonds are broken, a puff of methane gas is lost to space. A patch of ammonia ice would be lost as the comet crossed the orbit of Saturn. Carbon dioxide ice would begin vaporizing in earnest somewhere between the orbits of Saturn and Jupiter. Ordinary water ice would not begin significantly to vaporize until the comet neared the asteroid belt, between the orbits of Jupiter and Mars.

But it is precisely in the vicinity of the asteroid belt that most cometary tails are observed to form. This, in itself, is compelling evidence that water ice is a principal volatile constituent of the comets, and that water vapour and its degradation products are major components of the comas and tails of comets.

Occasionally, we do see comets release gasses, forming at least temporary comas and even tails when they are exterior to the asteroid belt. It is tempting to attribute such activity to the vaporization of other kinds of ices. But the further from the Sun that these outbursts occur, the less likely it is that they will be noticed by the handful of astronomers on Earth. Imagine a comet with a surface and interior composed of patches of different sorts of ices – water ice, methane ice, ammonia ice, carbon dioxide ice – and in an orbit that at first brings it only to the vicinity of Uranus, say, or Neptune. Each perihelion passage it will release methane ice as gas (and, if any, nitrogen and carbon monoxide ices). But these puffs of gas, even if on a significant scale, will remain unnoticed and unrecorded on Earth. After many perihelion passages the methane – at least

6.10 *States of matter in the present solar system at various distances from the Sun. Concentric orbits are shown – counting outward from the Sun – of the planets Mercury, Venus, Earth, Mars, Jupiter and Saturn. For three kinds of materials – silicates, water, and methane – the region in the solar system is shown in which they are gaseous (cloud), liquid (droplet), and solid (cube). Silicates are vaporized interior to the orbit of Mercury, water is liquid in the vicinity of the Earth, and frozen at greater distances from the Sun than Mars. Methane can be liquid in the vicinity of the orbit of Saturn, but is solid at great distances from the Sun. Thus, a comet made up of a mix of methane, water, and silicates and entering the solar system from afar has its methane becoming liquid and gas within the orbit of Saturn, its water within the orbit of Mars, and its silicates within the orbit of Mercury. (Painting by Jon Lomberg.)*

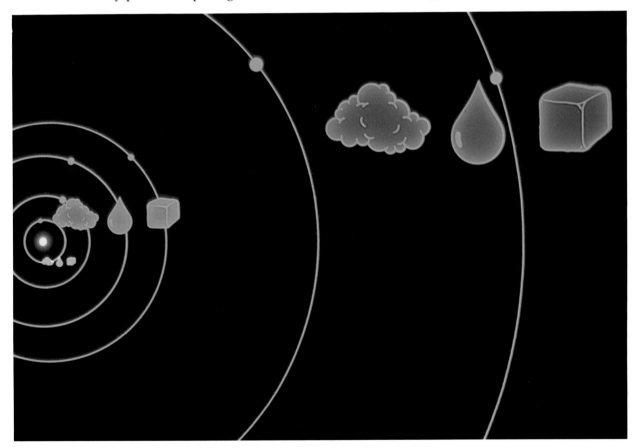

in the outer layers of the cometary nucleus – will have been entirely lost to space. The comet will have become devolatilized in methane. Each time it loses methane, it becomes, relatively speaking, richer in water. If such a comet is now perturbed – by a close approach with the planet Neptune, say – into the inner solar system, it will lose ammonia, if it has any, when passing Saturn. But it will be most easily detectable when it comes within the orbit of Mars, and heats up sufficiently for the strong lattice structure of ice to become disrupted. At this stage, an enormous cloud of the abundant volatile water comes pouring out. Thus, the prominence of water ice in comets is due to three factors: (1) the high cosmic abundance of water; (2) the possibility of loss of other volatiles in earlier incarnations of the comet's orbit; (3) the fact that the water comes spewing off only in the inner solar system when the comet is close enough for Earthbound observers to see.

Whipple and others showed that if a comet was made of ices, it would be able to supply copious quantities of molecules and small particles to form the coma and the tail. A cometary nucleus might lose a metre of material or more during each perihelion passage. If it started out with a radius of one kilometre, its substance would all be spent after a thousand perihelion passages. All that would be left would be materials like the involatile mineral grains – some of which, sooner or later, might be swept up by the Earth to produce a meteor shower. Think of two comets, both with perihelia near the Earth's orbit, but with different aphelia and different periods of revolution around the Sun. Suppose one of them is a short-period comet and takes five years per orbit. From the moment it arrives in the inner solar system to the moment it is entirely vaporized and converted into meteors is five thousand years. The long-period comet, with a period of a hundred years, say, takes instead a hundred thousand years to lose its volatiles. These are typical lifetimes for comets, provided there is a fresh ice surface exposed directly to sunlight during each perihelion passage. The layers of the cometary onion peel off once each orbit near every perihelion until eventually there is nothing left. Therefore, the population of short-period comets must be resupplied from a more distant repository of comets, just as Laplace and others had calculated (see Chapter 5).

Whipple also realized that his dirty ice model could, in a very natural way, explain the strange non-gravitational motion of some comets. Take an ordinary air-filled toy balloon, tie the nozzle closed, and let it fall. Its downward trajectory is slow and steady. But if you take the same balloon and hold it closed above your head, the nozzle squeezed

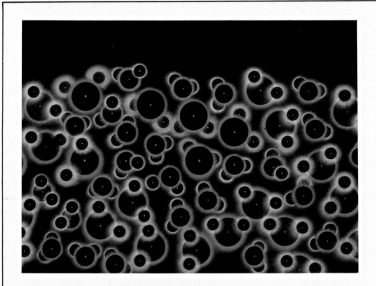

6.11 *Molecular structure of liquid water. The water molecules are not arranged in a silicate crystal lattice, but are free to move and are oriented at random. (Painting by Jon Lomberg.)*

LIQUID WATER

The kind of water ice we have described is the most common kind – those hexagonal crystal lattices characterize snowflakes and icebergs, glaciers and street slush on Earth. The structure is really quite remarkable. Ice has a lower density than liquid water because of the large void volumes or holes in the structure. This is why ice floats on liquid water. This is also why water expands on freezing. In the liquid (see Figure 6.11, above), the water molecules are moving too vigorously to have settled down into the regular ice lattice. They are busy tumbling and colliding into each other, and leaving little in the way of interior voids. But when the temperatures fall, the motion becomes less vigorous and the chemical bonds between adjacent water molecules assert themselves. The hexagonal lattice is formed and a material significantly less dense is formed. The density of liquid water is 1.0 grams per cubic centimetre, but the density of ordinary ice is 0.9 grams per cubic centimetre, a circumstance otherwise virtually unknown in nature. For other substances, almost always the solid is more dense than the liquid.

To prevent ice from evaporating directly into vapour, to detain it in an intermediary liquid stage, requires an atmosphere. Collisions with the overlying gas molecules then inhibit evaporation. With no atmosphere, there are no collisions, no such inhibition, and therefore no liquids. Even Mars, with an atmospheric pressure thousands of times that at the surface of a comet, cannot today maintain open bodies of liquid water. There can be no liquid water on a cometary nucleus because, even when its coma is most prominent, a cometary nucleus is still, by Earth standards, surrounded by high vacuum. You could imagine an interior pocket of the comet in which, if the temperatures were high enough, liquid water could readily form. But if the only source of heat is the sunlight from without, the temperatures will get colder the deeper inside the comet you are. Modern speculation about life in comets, discussed later, is closely tied to the possibility of liquid water deep inside the cometary nucleus, perhaps even oceans of underground water.

between thumb and forefinger, when you let it go, it briefly darts across the room making sudden starts and turns and occasional rude noises. This is the rocket effect. When the air rushes out of the nozzle, the balloon darts in the opposite direction. The reason is called Newton's third law of motion: for every action there is an equal and opposite reaction. A rocket works on exactly the same principle: the exhaust blasts down on the launch pad and the rocket lifts up into the sky. The rocket exhaust does not push against the ground; rockets work equally well – indeed better – in the vacuum of space. Another familiar example is the recoil of a rifle – the bullet goes forward, and the stock drives back into your shoulder.

Akin to the air in the balloon, the fuel in the rocket, and the bullet in the gun, is the ice in the comet. Imagine an iceberg tumbling towards the Sun, its surface covered with patches of rocky or organic material, interspersed with the more volatile ices. The surface of the comet is heated on approach to the Sun, and some of the ice warms up and vaporizes. You can imagine a little gush of methane or ammonia gas shooting into space – perhaps uncovering some deeper vein of the material, or perhaps only some rocky matrix material. But these jets of gas do not arise uniformly from all over the surface of the cometary nucleus at once. As a patch of methane ice vaporizes (action) at the orbital distance of Neptune, the orbit of the comet shudders slightly (reaction). Closer to the Sun, patches of ammonia or carbon dioxide can produce similar rocket effects. Detailed studies show that the tiny amount of non-gravitational motion in the short-period comets can readily be explained by the rocket effect from subliming veins of water ice on the surface of the comet.

It is warmest in the afternoon on Earth, rather than exactly at noon when the Sun is highest, because it takes a little time for the ground to warm up. The same is true on the comet, and it is therefore the afternoon side that heats up most and drives the bulk of the vaporized ice into space. Depending on which direction the comet rotates, this will either accelerate or decelerate its orbital motion.

Sometimes a comet is structurally weakened – by collision, or by rapid rotation, so it is straining against the cohesive forces – and it splits into two or more pieces. The new frozen volatiles now exposed to sunlight produce still further jets of gas and dust, and some additional small tumbling and darting of the cometary fragments.

Something similar may even occur in comets that never approach the Sun closely. The most famous case is Comet Schwassmann-Wachmann 1. It lives between the orbits of Jupiter and Saturn and undergoes episodic outbursts, occasionally brightening by a thousand times in just a few days. When there are no such outbursts it appears to be a dark, reddish object, quite similar to asteroids rich in organic matter. One view of the outbursts of Schwassmann-Wachmann 1 is that there are deeply buried repositories of exotic ices, being gradually heated by the Sun. Sunlight eventually turns a pocket of ice volatile, and water vapour gushes off the surface. Tiny grains of ice and dust produce a temporary cloud around the object – a distant coma – that we see only as a momentary brightening. The circular orbit of this comet speaks to a long residence time in this part of the solar system. So why are there any exotic ices still left near the surface? Perhaps this is not a case of exotic ices, but of something else – collision of the comet with boulder-sized objects in its orbit. The true explanation of the outbursts of Schwassmann-Wachmann 1 remains a mystery. It makes us wonder about other objects in the outer solar system – for example, Chiron, a small world that lives between Saturn and Uranus, or the icy moons of the giant planets. Might one of them, someday, abruptly wrap itself in clouds and brighten dramatically?

When the German mathematician F.W. Bessel witnessed the jets issuing from the nucleus of Halley's Comet in 1835, he speculated that a small, non-gravitational motion of the comet might result. The suggestion languished for 115 years, until Whipple revived it in a modern context. We can today see jets of matter spraying from the nucleus of a warm comet, irregularly turning on and off, like the attitude control jets of an interplanetary spacecraft. Every time one of these great fountains goes shooting off into space, the cometary orbit makes a little turn. It is even possible, as with Comet Swift-Tuttle (see Figure 6.13) to map the positions of the jets on the cometary nucleus.*

Thus, the dirty ice model in one swoop explains the timing and extent of the formation of cometary comas and tails, the non-gravitational motion, and the explosive jetting from cometary nuclei that were noted from at least the time of Thomas Wright. As we shall see, there is now, as well, direct evidence for water ice as the principal constituent of the comets that we observe. Fred Whipple's 'obvious' explanation, in the very finest tradition of science, produces a windfall of accurate prediction from a modest investment of hypothesis.

So, it looks very much as though comets really *are* giant iceballs racing around the Sun. How much ice is that? Imagine that you have snow ploughs stationed all over the Earth

6.12 *A fountain of cometary matter jetting out from the nucleus of Comet Halley in 1910. Drawing made at the telescope of the Helwan Observatory, Egypt, 25 May 1910. (Courtesy, NASA.)*

*Periodic comet Swift-Tuttle, 1862 III, is the source of the Perseid meteor stream. The jets mapped in Figure[6.13]are propelling future Perseid meteors into nearby space where, in time, they will fill much of the orbit of the comet, as discussed in Chapter 13.

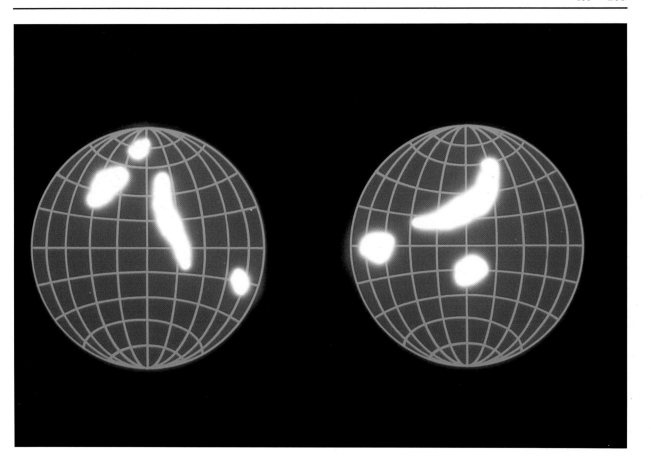

whose function it is to sweep up every flake of snow that falls in a year. And then imagine that all this snow is somehow packed into rough spheres, carried out into space, and put into cold storage. You would then have something like a cometary nucleus ten kilometres across. You could make a hundred ordinary nuclei with this much snow. Put another way, a typical cometary nucleus contains about as much snow as falls each year in Eastern Europe, or the Northern United States. It does not seem so very much.

In a way, Whipple's snowball hypothesis is a disappointment. Newton and Halley converted comets from a terrifying omen to a commonplace of nature, obedient to a divine mandate, unseen but manifest. But if comets are merely snowballs in orbit, is there any mystery left? Or are comets now demoted, become prosaic to the point of tedium? Remarkably, the new understanding of comets suggests that they are the key to the origin of the solar system, and to the surface characteristics of most of the worlds we know.

6.13 *(Above) Positions of jets on the nucleus of Comet Swift-Tuttle. The comet is here imagined to be spherical, and the bright, white regions show locales of jetting near perihelion as determined by Zdenek Sekanina of the Jet Propulsion Laboratory. (Painting by Jon Lomberg.)*

6.14 *(Opposite) Dirty ice comprising a cometary nucleus. (Painting by Michael Carroll.)*

COMET RAG

By

ED. C. MAHONY

Published by
Daly
MUSIC PUBLISHER
BOSTON, MASS

STARMER

Chapter 7

THE ANATOMY OF COMETS

There was another sign seen in Heaven,
and behold a great red dragon...
And his tail draweth a third part of the
stars in Heaven.

The Revelation of St John the Divine,
12. 14.

In all of human history, less than a thousand individual comets have been recorded, although a few hundred have been seen in more than one apparition, or passage by the Sun and Earth. Pliny wrote that no comets could be visible to the naked eye more than six months, which is still generally true today. In most cases, the comet is faint, barely visible as a bright smudge in the sky – a wisp of light, seemingly a small fragment of the Milky Way broken off and on its own. The visibility of such comets greatly improves with a small telescope, or even a pair of binoculars.

Comets do not streak across the sky; they rise and set with the stars. The notion of streaking stems partly from a confusion with meteors, and partly from the immediate impression given by drawings and photographs: in our everyday experience, objects with such shapes are usually streaking. When we see a picture of a comet, some of us are immediately reminded of a woman with long, straight hair being blown back behind her – the reason, as we have said, for the very name comet derived from the Greek word for hair. But the comet does not live down here, where there is air to blow things back. The comet lives in the nearly perfect vacuum of interplanetary space. And the tail does not always trail behind; instead, after perihelion passage and as the comet leaves the Sun, the tail precedes the nucleus. Something else is going on.

As a comet approaches the Earth, it typically increases both in its brightness, and in the apparent length of its tail across the sky. It may disappear into the Sun's glare as it negotiates perihelion passage, and then once more is visible – either brighter or dimmer, depending on the relative geometry of Sun, Earth, and comet. The tails of some comets visible to the naked eye stretch from zenith to the horizon.

7.1 *(Opposite) An Earthbound astronomer puzzles out the anatomy of a comet on this sheet music cover of 1910. (Courtesy, Library of Congress.)*

7.2 *The Great Daylight Comet of 1910. (From Henry Norris Russell. The Origin of the Solar System, Princeton University Press, 1935.)*

Somewhere around the orbit of Jupiter, a bright comet may first appear to the naked eye as a point of light – a star of fourth or fifth magnitude, say, surrounded by a perceptible haze. Comets vary greatly in brightness, of course; most are visible only through large telescopes; some can be seen with the naked eye; and occasionally – once every few years – there is one so prominent as to cause a general stir. About once a human lifetime, on average, a comet appears that can be seen in the daytime sky, even very close to the Sun. The Great Comet of 1910 (1910 I) – which in people's memories is sometimes confused with Halley's Comet, which appeared later that year – was such, and was called the Great Daylight Comet.

Most comets are found by astronomers, professional or amateur. Sometimes, during a total eclipse, a comet is discovered near the Sun, bathed in the light of the solar corona – a comet previously unknown, hitherto invisible, bleached out in the glare of the Sun. After the eclipse, the comet becomes invisible again. But such cases are rare*, and comets are most often discovered far from the Sun: time exposure of a field of sky brings out some faint, perhaps nebulous, object that is not on the standard charts. Amateurs – who can become passionately taken up with the search for new comets – sometimes systematically scan the heavens in strips, using special telescopes able to see large patches of the sky in a single view. Some amateurs have found more than a dozen separate comets over a lifetime of dedicated searching.

With atmospheric pollution and city lights, the practice is now becoming less fashionable, but once there were many people who knew the map of the sky like the back of their hands, and who could tell at a glance – when out on a brisk post-prandial constitutional – that there was a new point or smudge of light in the sky where no star had been before. Sometimes, exploding stars, novas, are discovered in this way; and sometimes comets. Occasionally, a bright comet is first detected with the naked eye by observers who are not astronomers at all. The classic case is the Great Daylight Comet of January, 1910, which was discovered by three labourers on the South African railroads. There were then few astronomers regularly scanning the skies of the Southern Hemisphere.

Every night, on average, somewhere on Earth, there is at least one astronomer peering through the telescope at a comet. This is seldom a part of the discovery process, but usually a segment of a much more intricate research

*Although it is possible to accomplish such an eclipse artificially, from spacecraft altitudes, by using an opaque disc to block out the Sun.

programme to understand the nature of comets. An astronomer might be photographing the comet, or passing its light into a spectrometer to uncover something about its composition and motions, or measuring the heat given off by the cometary nucleus. Most often, the discovery was not made by a human eye at the eyepiece of a telescope, but from a photograph of the sky, taken through the telescope for quite another purpose, which has been developed.

What do you look for when you think you've discovered a comet far from the Sun? You see a fuzzy patch of light. Might it be an artefact of the photographic emulsion? Take another picture. Could it be some other kind of faint patch of light – a nebula or a distant galaxy – not listed on your chart of the sky? Take another picture. If it moves with respect to the stars at reasonable speeds, then you have probably discovered a comet. If it isn't fuzzy, it might be a new asteroid.

Discoveries of comets by professional astronomers are mainly made when the comet is far from the Sun, either on approach or – more rarely – after perihelion passage. Almost invariably (except when a known comet is being 'recovered') such discoveries are made as an incidental by-product of some quite different study. Discoveries by amateur astronomers, in contrast, are most often made when the comet is very close to the Sun, that is, within a few hours of sunset or dawn. Because comets may brighten erratically as they throw off successive comas, a comet which had been invisible to a large astronomical telescope when it was far from the Sun may sometimes be discovered with a very modest instrument, or even with the naked eye, when it is close to the Sun – as was the case with the Great Daylight Comet of 1910.

A good fraction of comets with orbits reliably established as periodic are 'recovered' – seen on some future approach to the Earth. But how can you be sure that what you are seeing is the same comet observed years ago? The comet generally has no distinctive identifying characteristics – no regimental tie, no tartan. Nevertheless, there are ways. Following Edmond Halley's pioneering work, the orbital characteristics of the earlier comet and the present one are compared: the period, eccentricity, distance of perihelion from the Sun, and inclination of the orbit, for example. For the 1986 return of Halley's Comet, astronomers had calculated precisely where the comet should be at every point in its orbit. It was detected with large telescopes more than a year before perihelion, when it was in the vicinity of the orbit of Saturn.

Discoveries of comets are sometimes made that remain unconfirmed – usually because the orbital periods are undetermined or too long. Occasionally, short-period comets with well-determined orbits are not recovered in their next

7.3 *Stamps from the Soviet Union, Chile and Belgium commemorating comets.*

*(Opposite) The components of a comet. A typical cometary nucleus, **7.4** (top), a few kilometres across, is shown for scale at the diffuse outer edge of a typical cometary coma. A coma in the inner solar system is shown, **7.5** (inset left) next to an image of the Earth. The coma is composed of diffuse gas and fine particles, with the nucleus being an insignificant and in this case invisible dot at the centre. A well-developed cometary tail, **7.6** (inset right) extends from the orbit of Earth to the orbit of Mars. (Diagrams by Jon Lomberg/BPS.)*

NAMING THE COMETS

Comets are often named after their discoverers, as Comet Ikeya-Seki, or Comet West. There is even a Purple Mountain Observatory Comet, discovered in a time when individual achievement was out of favour in the People's Republic of China.

Sometimes, comets are named not after their discoverers, but after those who first recognized that the comets seen in two or more apparitions were really the return of the same comet. Halley's Comet is such a case, as is Encke's.

Comets are also designated by a Roman numeral, designating position in the sequence of perihelion passages in a given year – as 1858 VI, or 1988 I.

apparition. Many such reports are probably not due to an error made by the astronomer, but to a comet, ordinarily too faint to see, undergoing an explosive outburst of gas and dust. When the activity is over, the comet returns to obscurity. These unrecovered comets serve to remind us that there is a vast population of undiscovered comets.

In seeking to understand what comets are, where they come from and what significance they conceivably might hold for us, we are the beneficiaries of thousands of years of patient observation and recording of data by people all over the planet. An official scribe of Han Dynasty China or Seleucid Babylon, matter-of-factly noting the properties of a comet, helps us millennia later to test Newtonian gravitational theory, or fluctuations in the arrival rate of new comets. We cannot, of course, thank that scribe; he could never know that astronomers in the far future would use his observations to draw conclusions that would astonish him. But it is well for us to remember our debt to all those who have patiently built, brick by brick, this edifice of knowledge.

From the cumulative weight of astronomical observation, an overall picture of the nature, the anatomy of comets has emerged. We have mentioned various components of this picture already; let us now take stock.

Comets are by far the largest and most variable objects visible in the solar system.* A tiny nucleus produces a substantial coma and an enormous tail, often much larger than

---

*Charged particles in Jupiter's magnetotail – formed by the interaction of the solar wind with Jupiter's magnetic field – reach to the orbit of Saturn; this planetary magnetotail is larger than any known comet tail. But it was discovered only when a spacecraft flew through it.

the Sun. But for two comets at the same distance from the Sun, one may have an immense tail, and the other none at all. Following the spirit and precedent of Thomas Wright (see Chapter 4), we show the relative scales of a typical cometary nucleus, coma, and tail in Figures 7.4-6. When it can be seen at all, the nucleus of a comet appears through the telescope as a starlike point of light. Generally it is a few kilometres across, and yet this tiny ball of ice can generate a visible tail that is longer than the distance between the orbits of the planets in the inner solar system. A one kilometre object with a tail a hundred million kilometres long is like a speck of dust dancing in the sunlight in Washington DC, with a tail that reaches Baltimore.

In the earlier scientific literature there are reports of direct measurements of the nuclei of comets (see Figure 7.7), generally reported to be hundreds or thousands of kilometres in diameter. It now appears that these measurements – made near the closest approach of the comet to the Earth – were in fact of the brightest part of the coma; the nucleus itself, much smaller, must have been lying hidden inside the coma. A few observers have claimed that they could see a background star winking off and on as the nucleus passed in front of it.

Sometimes, by chance, a comet travels through the line of sight from the Earth to the Sun. If the nucleus were huge – 1,000 kilometres across, say – it might appear as a small, moving, black dot silhouetted against the solar disc. But in

7.7 *The nucleus of the Comet of 1618 (top), drawn at the telescope on three different dates by Cysatus, and the Comet of 27 December 1652 as drawn by Hevelius. What these observers, working with the earliest astronomical telescopes, actually saw were the comas, not the nuclei of the comets in question; the details within the coma are probably spurious, caused by the imperfections of the seventeenth-century lenses. (From Amédée Guillemin,* The World of Comets, *Paris, 1877.)*

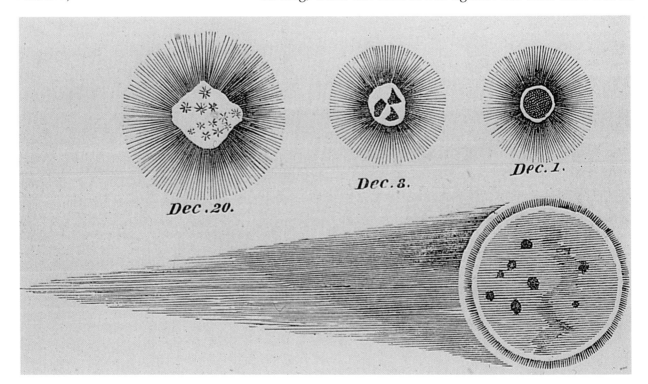

every case where the observations have been attempted – as, for example, when the Great Comet of 1882 transited the Sun – the nucleus was too small for any trace of it to be seen.

When the comets come close to the Earth they veil themselves from view by inquisitive astronomers. In general, they only undress in the dark – beyond the orbit of Jupiter.

Occasionally, a comet comes close to the Earth without a thick coma, and a small, bright point within the coma can be seen. This may be the cometary nucleus, but it might also be a dense inner coma of dust.

If you can catch the comet when it is far from the Sun, before it has generated a coma, then you might catch a glimpse of the naked nucleus. But because comets are so small, when they are at great distances from the Sun we can only see them as points of light no matter how bright they are, and no surface detail is discernible. There is an intrinsic limit to the resolution of ground-based telescopes, and to our ability to distinguish fine detail. So astronomers have devised another way. The amount of sunlight reflected back from the nucleus to the Earth is measured: then, if only we knew how dark the nucleus is, we could calculate how big the nucleus has to be to return the amount of sunlight we measure. The more reflective the surface, the smaller the nucleus must be to return the measured amount of light; conversely, the darker it is, the larger it must be. For dark, dirty cometary nuclei, the calculated sizes are a few kilometres across, or less.

The most direct way to determine the size of a cometary nucleus is to send a burst of radio waves out from the Earth to intercept the comet and be reflected back. Even a dense planetary atmosphere, like that on Venus, is transparent to such radar probing; these radio waves will pass through the coma with no difficulty at all. The few comets that have come close enough to Earth since the advent of advanced radar astronomy show nuclei between one and a few kilometres across, in good agreement with the best estimates made by other techniques.

A cometary nucleus five kilometres across, roughly the size of Halley's Comet, doesn't sound like much; but if it were gently placed at a typical locale on the Earth's ocean beds (see Figure 7.8), it would stick up smartly above the ocean surface, making – at least until it melted – a novel kind of tropical island.

The number of comets that have been so measured represents less than one percent of the known comets. How likely is it that much smaller or much larger comets exist? A comet a few kilometres across is the size of a city. Are there then comets as small as a house, or ones as large as Luxembourg, say, or Brunei? Very small comets in the inner

7.8 *Schematic representation of a cometary nucleus five kilometres across, roughly the size of Halley's Comet, that has come gently to rest at a typical locale on the bed of one of Earth's oceans; a considerable part of the nucleus sticks up above the surface of the water. (Diagram by Jon Lomberg/BPS.)*

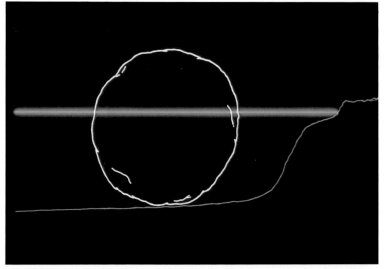

7.9 *(Below) Through violent jetting a comet fragments into many small pieces. The vaporization of ice destroys the comet. (Painting by William K. Hartmann.)*

solar system might be impossible to detect; they also would have an extremely brief lifetime, as their ices vaporized and the comet dissipated. As comets split and break up, there are surely fragments released that are the size of a house or smaller, but they do not survive for long. Oddly, astronomers do not find more and more small comets by looking at fainter and fainter objects. There appears to be a real lack of comets smaller than a football field (a hundred metres across). No one knows why.

There is indirect evidence of occasional long-period comets that are considerably larger, perhaps a hundred kilometres across or even more. The most celebrated example is the Great Comet of 1729. It was visible to the naked eye, although its perihelion was in the exterior of the asteroid belt, almost at the orbit of Jupiter. If it had instead come all the way to the vicinity of the Earth, it would have been bright enough to read a newspaper by at midnight.

For it to have been so bright so far away, it must have both been very large and releasing great quantities of exotic ices. In the vicinity of Jupiter, it is much too cold for significant vaporization of ordinary water ice. As we have seen, an iceberg of frozen nitrogen or carbon monoxide begins vigorous evaporation well beyond the orbit of Pluto; and one of methane just a little inside the orbit of Pluto. Any comets made mostly of such materials would have spent much of their substance before ever coming close enough to Earth to be seen. Comets made of ammonia or carbon dioxide would vigorously evaporate and dissipate between the orbits of Jupiter and Saturn. But unless such comets are particularly massive, their outgassing and jetting should remain indetectable from Earth.

Comet Kohoutek was noticed to be exceptionally bright when it was still far from the Sun, from which widely-publicized conclusions were drawn about how brilliant it would be as it passed the Earth in December 1973. But the apparition was far from spectacular; indeed, Comet Kohoutek was just barely visible with the naked eye from the surface of the Earth, although it was clearly visible to the Skylab astronauts in the Earth's orbit. Apparently, it was so bright so far from the Sun because of the vaporization of exotic ices, a process which had been entirely completed by the time the comet came close to the Earth.

From the periodic appearance of jets, it has been possible to calculate the rotation period of the cometary nucleus sequestered inside the coma. The spin rates of dozens of comets have now been measured by this or other methods. A typical comet turns out to rotate once every fifteen hours – not too different from the length of the day on Earth. The direction

(Opposite) a family portrait of some of the irregular small moons in the solar system. Because their gravities are so weak, they are not compressed into a spherical shape. **7.10** (top left) Phobos, the innermost moon of Mars. **7.11** (top right) Deimos, the outermost moon of Mars. **7.12** (centre) Three views from different vantage points of the Saturnian moon, Hyperion. **7.13** (bottom) Some of the small moons of Jupiter and of Saturn. The shapes of cometary nuceli are expected to resemble these. (Courtesy, NASA. Photographs obtained by the Viking and Voyager missions.)

of the spin axis in the sky seems to be entirely random: cometary spin axes do not, for example, mainly point towards the North Star. For some comets, observations ranging back in time over a century or more can be used to determine rotation rates. Comet Encke, for example, has not changed its rotation period much in 140 years.

Because the typical comet is so small, its gravitational pull is tiny. If you were standing on the surface of a cometary nucleus, you would weigh as much as a lima bean does on Earth. You could readily leap tens of kilometres into the sky, and easily throw a snowball to escape velocity, as we imagined in Chapter 1. The Earth and the other planets tend to be almost perfect spheres because, as Newton showed, gravity is a central force – it pulls everything equally towards the centre of the world, itself held together by the force of gravity. The mountains sticking up above the spherical surface of the Earth represent less of a deviation from a perfect sphere than does the layer of paint or enamel on the surface of a typical globe that represents the Earth. If you were able to pile a sizeable mountain on top of Mount Everest, it would not just sit there poking in solitude into the stratosphere. The additional weight you had added would crush the base of Everest, and the new composite mountain would collapse until it was no larger than Everest is today. The Earth's gravity severely limits how much deviation from a perfect sphere our planet is permitted.

On a comet, on the other hand, the gravity is so small that odd, lumpy, potato-like figures would not be squeezed into a sphere. Such shapes are already known for the small moons of Mars, Jupiter, and Saturn. On a typical comet, you could build a tower reaching a million kilometres into space, and it would not be crushed by the comet's gravity – although it would certainly be flung off into space by the comet's rotation, which we discuss later. Both the small moons and the cometary nuclei stay lumpy because their gravities are low. But they *get* lumpy for different reasons – the moons because of the history of collision with other bodies, and the comets because they accreted irregularly, or because they have vaporized surface ices irregularly.

What is the pressure at the centre of a cometary nucleus? Pick one that has a radius of a kilometre. Everyone has the sense that the pressure from the weight of the overlying rocks one kilometre beneath the Earth's surface is considerable. We know about coal-mine shorings collapsing and miners being crushed. But if the gravity were thirty thousand times less, then the overlying rocks would weighh thirty thousand times less. Put another way, the pressure at the centre of such a comet is the same as it is at 1/30,000 of the comparable depth

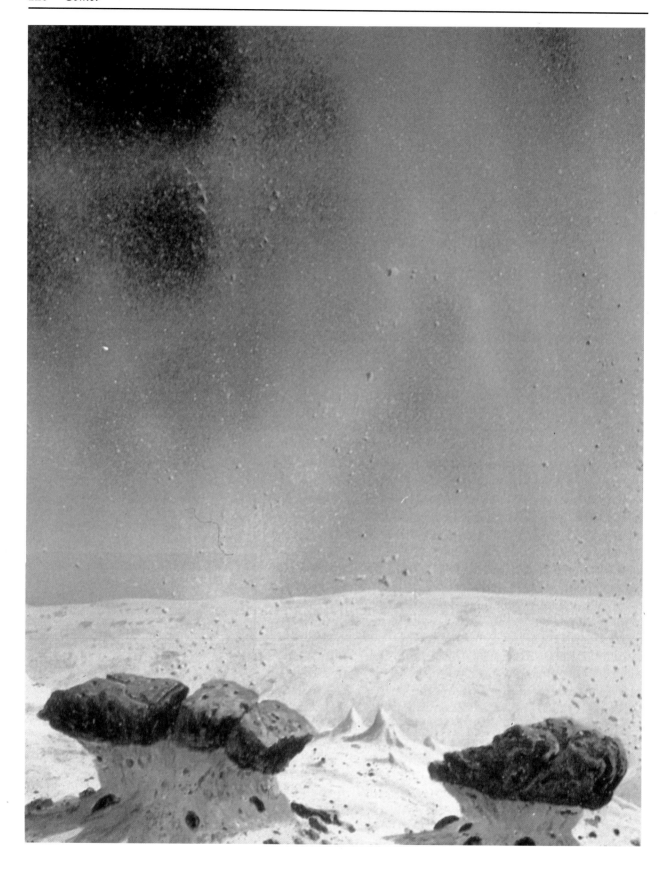

blowing into the interplanetary vacuum, and entraining cometary dust. Some comets are very dusty, others comparatively clean. Even in an aging cometary nucleus, there are still localized jets of gas and sources of fine dust. While unpredictable, there is a tendency for both jetting and non-gravitational forces to increase the closer the comet is to the Sun, consistent with the vaporization of ices and the rocket effect.

Dusty comets have been observed to pour tons of fine particles into interplanetary space every second, and for most comets several times more water is lost than solids. Besides tiny particles of dust swept off into space in the jets, there may also be fragile clumps of dust. A cloud of particles at least centimetres across surrounding the cometary nucleus has been found by radar. The relative amount of dust and ice near the surface of a cometary nucleus is probably variable from comet to comet.

Comet Schwassmann-Wachmann 1, orbiting the Sun always beyond Jupiter, has produced over a hundred known outbursts, an average of about two a year. In successive perihelion passages, some comets progressively lose layer after layer of volatiles, until there is nothing left. Others seem to lose a metre or so of ices each perihelion passage for a number of orbits, and then close down shop. It may be that the refractory, rocky stuff that has not gushed away to space with the ices is now abundant on the surface of the nucleus, preventing sunlight from penetrating to the cold interior, and buried ices, even if warmed, from making their way to the surface. After successive evaporations, the surface of a comet may resemble the lag deposit on a glacier, the involatile rocky material covering a deep layer of still frozen ice. An important unsolved problem in planetary science is whether such end products of cometary evolution – rocky mantles surrounding icy cores – corresponds to any known worlds in the solar system.

If we include the long tails of the great comets, these objects are larger than anything else seen in the solar system – larger than the planets, larger even than the Sun and its corona. Indeed, except for the red giants, comets are larger than any known star. The longest comet tails are greater than the distance from the Earth to the Sun. Even at its widest, the width of the fan-shaped tail is no more than a few percent of its length. The material in the coma and tail of a comet never returns to the nucleus, is never recaptured by the comet. Thus, as time goes on, the comet gradually dwindles, successive layers are peeled off and lost to space, and its interior parts are exposed to view. Every comet we can see is dying.

7.18 (Opposite) *Explosive jetting on the surface of a cometary nucleus. The evaporation of ice during successive perihelion passages leaves a lag deposit of rocky material behind. (Painting by William K. Hartmann.)*

We think ourselves unhappy when a comet appears, but the misfortune is the comet's.

Bernard de Fontenelle,
The Plurality of Worlds
(Paris, 1686).

Chapter 8

POISON GAS AND ORGANIC MATTER

Hast thou ne'er seen the comet's flaming flight?
The illustrious stranger passing, terror sheds
On gazing nations from his fiery train
Of length enormous; takes his ample round
Through depths of ether; coasts unnumber'd worlds
Of no more than solar glory; doubles wide
Heaven's mighty cape; and then revisits earth,
From the long travel of a thousand years . . .

Edward Young,
Night Thoughts,
(1741).

William Huggins scared the world. It was entirely unintentional. No one could have foreseen it. He was minding his own business, which happened to be astronomy. But because of Huggins, in 1910 there was a sustained national panic in Japan and Russia lasting for weeks; a hundred thousand people in their nightclothes filled the rooftops of Constantinople; Chicago apartment dwellers anxiously stuffed rags under their doors; and Pope Pius X condemned the hoarding of cylinders of oxygen in Rome. A dispatch from Lexington, Kentucky – typical of reports from all over the world – announced that 'excited people are tonight holding all-night services, praying and singing to prepare themselves ... and meet their doom'. Through fear of the forthcoming catastrophe, a smattering of people in many lands took their own lives. But this is getting ahead of the story.

Huggins was one of the first astronomical spectroscopists – scientists who decompose light into its constituent colours, or frequencies, and are able to deduce the motion and composition of a distant object. Spectroscopy traces back to another towering accomplishment of Isaac Newton. By passing a chink of sunlight, admitted into a darkened room, through a prism of glass, Newton showed that ordinary white light is really a mixture of light of many colours. Different colours of light were bent by different angles when passing through the prism, so they could be spread out against a surface on the other side of the prism. At first, the surface was something like a white piece of cardboard. Much later, it

8.1 *(Opposite) The Earth and Moon pass through a comet tail, with the Milky Way in the background. (Painting by Rick Sternbach.)*

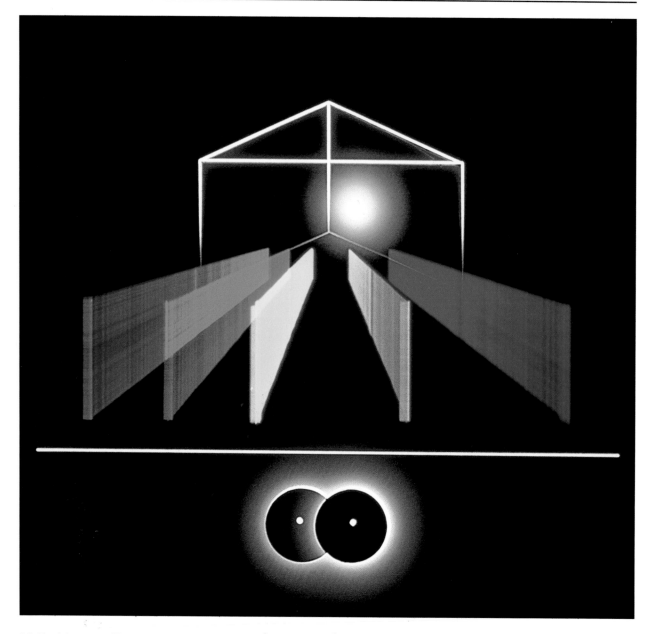

8.2 *The light given off by a comet enters a telescope and passes through a prism (rear), where its constituent frequencies are separated out producing a spectrum. The spectrum schematically indicated here is that of C_2 – the molecule composed of two carbon atoms, symbolized at the bottom. (Diagram by Jon Lomberg/BPS.)*

became a photographic emulsion. The machine built around the prism was called a spectrometer, and the rainbow pattern produced was called a spectrum (plural, spectra).

When a spectrometer of higher dispersion was used, it turned out that sunlight was composed, in addition to its rainbow continuum, of a set of irregularly spaced dark lines, representing missing frequencies. It was soon determined that these lines are produced when light from the hotter, deeper layers of the Sun is absorbed by a cooler higher atmosphere of gas surrounding the Sun. Each chemical element absorbs a different set of frequencies and produces a different set of dark lines. In the laboratory, when a spectrum of a mix of chemical elements (and simple molecules) was obtained, the dark lines of all the constituent atoms would be seen. Soon it became possible to identify some of the spectral lines in sunlight, by comparison with laboratory experiments on the absorption spectra of common materials on Earth.

Each chemical element left its own unique signature in the spectrum. You could accumulate a catalogue of such signatures and then recognize, in the spectrum of the Sun, evidence of its constituent atoms. Scientists found themselves measuring, to their own astonishment, the constitution of the Sun and stars. Spectroscopy revolutionized all the science of the time and astronomy in particular.

Huggins was the beneficiary of a spectroscopic technology just reaching maturity. He put his spectrometer at the focus of large telescopes and looked at everything he could. He was the first to show that the stars were made of the same chemical elements as the Earth and the Sun. He demonstrated Edmond Halley's conjecture that certain interstellar nebulae were vast clouds of glowing gas. When Comet Winnecke passed near the Earth in 1868, it was a matter of course for Huggins to examine the spectrum of its coma. He confirmed a finding made for another comet by Donati in 1864 that there were three *bright* bands in the blue part of the cometary spectrum. Huggins also found that in the spectra of the comets there were two components – a continuum of colour interspersed with dark absorption lines that he correctly guessed was merely reflected sunlight, and the three bright bands that had been seen by Donati. If a dark line meant that some atom or molecule was absorbing light, then a bright line, or a set of them, meant that something was emitting light. But what?

The only way to answer this question was to look at the emission spectra of a wide variety of materials in the laboratory. In 1868 Huggins made a surprising discovery. He sparked ethylene (C_2H_4) – something like the natural gas used in household ovens – examined it in the spectrometer, and found exactly the same three bright bands as had been detected in emission from the comets. He concluded:

> There could be no longer any doubt of the oneness of chemical nature of the cometary stuff with the gas we were using, in fact, that carbon, in some form or in some state of combination, existed in the cometary matter.

The three bright bands are now known to be produced by the molecular fragment C_2, two carbon atoms bonded together.

When the Great Comet of 1881 appeared, Huggins again obtained a spectrum from which were deduced not only the presence of C_2, but also C_3, CH, and CN – a rich harvest from the first cometary spectrum to be photographed. (Huggins himself essentially identified C_2 and CN, and argued for

8.3 *A cometary spectrum by William Huggins, showing dark absorption lines (K, H, b, G, etc) superimposed on three clusters of bright lines representing emission from the comet. (From* Proceedings of The Royal Institution, Volume 10.*)*

8.4 *Nineteenth-century spectra of emission bands in (2) olive oil, (3) ethylene, and (4) and (5) in two comets. Comet Winnecke II seems to have more in common with olive oil than it does with Comet Brorsen I. All the comparison actually shows is that olive oil, ethylene and cometary nuclei, when heated or irradiated, give off molecules of C_2. (Diagram from Camille Flammarion,* Astronomie Populaire, *1880.)*

8.5 *A modern cometary spectrum. Comet Kobayashi-Berger-Milon shows very prominent spectral features due to C_2 and CN; the spectral signatures of other molecular fragments are also shown. At the bottom is a scale of wavelengths that runs through the visible spectrum. Spectrum from the Wise Observatory, Tel Aviv. (Courtesy, NASA.)*

8.6 *The quite sedate title page of the introduction to the section on comets in Camille Flammarion's book* Astronomie Populaire *(Paris, 1880).*

hydrocarbons, of which the molecular fragment CH is the simplest exemplar.) We now know that it is hard to find a comet *without* the spectral features of C_2, C_3 and CN in its coma. Huggins was struck by the fact that the material in the comets was similar to complex organic matter of unquestioned biological origin on Earth. Many scientists cautiously concluded that the carbon compounds found by Huggins in the comas of comets were, as one of his contemporaries wrote, 'the result of the decomposition of organic bodies'. Only rarely voiced was the key question: Were such 'organic bodies' of biological origin?

These discoveries of Huggins and his successors interested virtually no one – until 1910, when it seemed the Earth was about to brush against the tail of Halley's Comet. The molecular fragment CN, a carbon atom attached to a nitrogen atom, had been detected in the comas and tails of many comets, and then confirmed in Halley's Comet as well. It was called cyanogen. But when chemically combined in a salt it had another name – cyanide. Since only a grain of potassium cyanide touched to the tongue is sufficient to produce instant death, the idea of the Earth flying through a cloud of cyanide produced an understandable apprehension. People imagined themselves, choking, gasping, and dying millions asphyxiated by the poison gas.

The global pandemonium about poison gas in Halley's tail was, sadly, fuelled by a few astronomers who should have known better. Camille Flammarion, a widely-known popularizer of astronomy, raised the possibility that 'the cyanogen gas would impregnate the atmosphere [of the Earth] and possibly snuff out all life on the planet'. With vaguely similar pronouncements by Gambert and Laplace preceding him, and a traditional dread of comets going back to prehistoric times, it is not surprising that statements like Flammarion's helped to produce a worldwide comet frenzy.

In fact, it was not even clear that the Earth would pass within the tail of Halley's Comet. In any case, the tails of comets are extraordinarily thin, a wisp of smoke in a vacuum. The cyanogen is, in turn, a minor constituent in the tails of comets. Even if the Earth *had* passed through the tail of Halley in 1910, and the molecules in the tail were thoroughly mixed down to the surface of the Earth, there would have been only one molecule of cyanogen in every trillion molecules of air – a good deal less than the pollution caused, even far from cities, by industrial and automobile exhaust (and much less than would occur in the burning of cities in a nuclear war). Also, the Earth, half a century earlier, had survived the comet of 1861 with no apparent ill-effects.

All this, or something like it, was made clear by the global

8.7 *Not all public interest in the 1910 apparition of Halley's Comet was connected with poisoned gas. (Courtesy, Library of Congress.)*

8.8 *French stamp commemorating the publication of Camille Flammarion's book* Astronomie Populaire.

astronomical community, but – like the panic in 1979 over the possibility that the Skylab satellite would fall on the heads of innocent passers by – the assurances had little effect. Why should the world's population be so susceptible? This was certainly a time of burgeoning industrial pollution and attendant respiratory disease. But perhaps, also, this fear was not so much about poison gas up there in the comet as down here in the national arsenals. In H.G. Wells' *The War of the Worlds* (1898), the Martians invading the Earth used gas warfare. This was one of many novels of the period depicting a future war in which devastating poison gas attacks were vividly portrayed. Far from being merely the inventions of science fiction authors, these works reflected a dismaying reality. At least until a few years before the 1910 apparition of Halley's Comet, new approaches to chemical warfare were

8.9 *Two views on successive nights in 1910 of Halley's Comet, when the Earth may have grazed the tail of the comet. (Courtesy, International Halley Watch.)*

8.10 *'Waiting for the End of the World', by R. Jerome Hill.* (Harper's Weekly, 14 May 1910.)

8.11 *Sir William Huggins (1824-1910).* (Courtesy, Mary Lea Shane Archives of the Lick Observatory, University of California.)

being actively pursued in European military establishments. At the Hague conference of 1899, proposals for banning certain chemical weapons were offered, but they were successfully opposed by John Hay, the American Secretary of State. A resolution forbidding asphyxiating gasses in artillery shells was adopted, but the US did not sign. When the First World War came, four years after the comet, 120,000 tons of poison gas were used, mainly by Germany, France and, to a lesser extent, the United States, producing one-and-a-quarter million casualties. More than a quarter of all American casualties in the First World War were due to poison gas. Only a small fraction of the poisons employed were cyanides.

But in 1910, the comet passed, no one was asphyxiated, and the Earth seemed none the worse for the experience. The most sensitive tests revealed not a trace of increased cyanogen in the air – a source of relief to many, although no comfort to the discoverer of cometary cyanogen. Less than a week before the closest approach of Halley's Comet to the Earth, Sir William Huggins, the man who frightened millions by minding his own business, died at the age of eighty-six.

Since then, compelling evidence has emerged for the presence of frozen water and silicate minerals in addition to organics in the comets. A list of the variety of molecules found to date is given in Figure 8.12. But most of these simple molecules and molecular fragments are not present on the cometary nucleus; rather, they are bits and pieces broken off from the comet and released into space by radiation from the Sun. They were originally parts of larger, so-called parent molecules, whose identities are conjectured, but which remain largely unknown.

In visible light, the spectra of comas tend to be dominated by the blue emission of C_2. This molecule is unfamiliar because it breaks up or combines with other molecules upon collision. If you had some in the air before you, it would not last long. But in a comet, the densities are so low that a long time will pass before C_2 collides with another molecule. Imagine a C_2 molecule sitting in space, exposed to radiation from the Sun. A photon of ultraviolet light, invisible to our eyes, hits it and raises it to what is called an excited state. It cannot get rid of this additional energy by collision, since there are no collisions. By nature, C_2 is inhibited about emitting a photon of ultraviolet light back into space. So it simply sits there in the coma, throbbing.

Eventually, it does spit out one or more photons – not in the ultraviolet, but rather in the visible part of the spectrum – where our eyes are sensitive and our atmosphere is transparent. If we irradiated a collection of C_2 molecules with ultraviolet light and then stuck them in the dark, they would begin to glow, a process called fluorescence. The gasses are induced to glow by the intense but invisible ultraviolet radiation from the Sun. Accordingly, since the dust and ice in the comet reflect or scatter sunlight, there are two sources of light from the comet – scattered sunlight and fluorescence. Other molecules may be much more weakly fluorescent, or may fluoresce in other wavelengths of light that are not so readily detectable from the surface of the Earth. So the prominence of certain molecules such as C_2 in the spectrum of comets does not necessarily mean they are very abundant in the comet. But their abundance is high enough to indicate that their parent molecules – organic molecules – must be reasonably common in the cometary nucleus from which the coma derives. Much more common is the molecular fragment OH, produced by tearing water (HOH = H_2O) apart.

Comets were not observed in the ultraviolet part of the spectrum until rockets and, later, orbiting observatories were aloft, above the Earth's atmosphere, in the beginning of the 1970s. These first observations showed that comets have an envelope of hydrogen gas that extends for millions of kilometres from the cometary nucleus away from the Sun. Accompanying the hydrogen is OH, the two together evidently produced by the dissociation of water making up the cometary nucleus. The identification of ionized water, H_2O^+ (a water molecule with a negatively-charged electron missing) in comets provides another link in the chain of evidence for water as a major constituent of the cometary nucleus. These results, both qualitatively and quantitatively provide powerful additional support for Whipple's dirty ice model of the cometary nucleus.

Subsequent ultraviolet observations by orbiting observatories, and by rockets specially fired above the Earth's atmosphere for the purpose, have revealed a number of new molecules and molecular fragments, including S and CS, the first discovery of sulphur and its chemical products in comets. The most likely parent molecule of CS is CS_2, and it, in turn, probably derives from a more complex sulphur-containing organic molecule. In most cases, the cometary spectra invite us to trace backwards from simple molecular species, that we can readily identify, to progressively more complex and more uncertain parent molecules, among which may be complex organic molecules of the sort that are found in the interstellar grains and gas, discussed below.

Infrared spectra of both the coma and the tail of comets (for example, Comet Kohoutek) show a spectroscopic emission feature due to silicates, the principal constituents of rocks.

Beginning with observations of the Great September Comet, 1882 II, the spectroscopic signature of metals has been

[The Laputans'] apprehensions arise from several changes they dread in the celestial bodies. For instance... that, the Earth very narrowly escaped a brush from the tail of the last comet, which would have infallibly reduced it to ashes; and that the next, which they have calculated for one and thirty years hence, will probably destroy us.

Jonathan Swift,
Gulliver's Travels (1726).

Key to figure 8.12
1) O
2) OH
3) OH^+
4) H_2O
5) H_2O^+
6) CO_2^+
7) CO^+
8) CO
9) CH
10) H
11) C
12) C^+
13) CH^+
14) NH
15) NH_2
16) N_2^+
17) CN
18) CN^+
19) CH_3CN
20) HCN
21) C_3
22) C_2
23) S
24) CS
25) silicates
26) Co
27) Cu
28) Fe
29) Ni
30) Mn
31) Ca
32) Ca^+
33) V
34) Na
35) K

Key to figure 8.13
1) H_2
2) CH
3) CH^+
4) OH
5) CO
6) CO^+
7) SiO
8) SiS
9) CS
10) NO
11) NS
12) H_2O
13) COH^+
14) SiH_4
15) NH_3
16) HN_2
17) CN
18) HCN
19) HCO
20) HCO^+
21) H_2S
22) HNC
23) CCH
24) OCS
25) SO_2
26) SO
27) HCS
28) $CCCN$
29) $HCCH$
30) HNO
31) $HNCS$
32) H_2CS
33) $CCCCH$
34) $CCCO$
35) $HCOOH$
36) HC_3N
37) HC_5N
38) HC_7N
39) HC_9N
40) $HC_{11}N$
41) CH_4
42) H_2CO
43) H_2CNH
44) H_2NCN
45) HCl
46) CH_3OH
47) CH_3CN
48) C_2H_4
49) C_2H_3CN
50) $HNCO$
51) CH_3SH
52) CH_3NH_2
53) CH_3CCH
54) $HCONH_2$
55) CH_3CHO
56) $HCOOCH_3$
57) CH_3CCCN
58) CH_3OCH_3
59) CH_3CH_2OH
60) CH_3CCCCH
61) CH_3CH_2CN

found in the sungrazing comets, comets that come very close to the sun. Such atoms as chromium, nickel, and copper can be detected, although they are never observed in other comets. The reason is clear: as Isaac Newton had first pointed out, sungrazers come so close to the Sun that even iron would become red hot. The individual mineral grains evaporate and boil, and a gas composed of metal atoms is poured off into the surrounding coma for astronomers on Earth to view. In the cold and dark from which the comets come, the metals are probably chemically bound to the silicates.

Even for comets that do not come very near the Sun, the constituent dust grains and organic molecules find themselves in a kind of blast furnace during perihelion passage – compared to the staid and safe environment they enjoy far from the Sun. High energy solar photons and charged particles are constantly bombarding cometary molecules, breaking them into pieces, tearing fragments off, ionizing, dissociating. Many of these molecules are so fragile that they last only hours or less in the ultraviolet radiation flux from the Sun before they are torn apart. This is another reason why the molecules observed in comas and tails are not necessarily the most abundant molecules, but merely the ones most resistant, most impervious, to the radiation that bombards them.

Since methane ice (CH_4) is expected to be abundant in the outer solar system, it is natural to consider it the parent molecule for such fragments as CH. But, unless a new vein of methane ice is by chance exposed just as the cometary nucleus enters the inner solar system, CH_4 methane ice is unlikely at such high temperatures; and for this reason, it is attractive to consider the methane trapped as a clathrate in the ice crystal lattice. However, no amount of methane clathrate can explain such molecules as C_2 and C_3, which must derive from more complicated organics. Thus, even the simplest cometary organic molecules, such as CH, probably derive from more complicated organics and not from methane.

In addition to measurements in the visible, ultraviolet and infrared parts of the spectrum, there is also spectroscopy at radio frequencies. It does not use prisms, but it is perfectly able to distinguish absorption or emission lines caused by particular atoms or molecules. So far as is known, the first attempt at radioastronomical observations of a comet was performed during the closest approach of Halley's Comet to the Earth on 18 May 1910, by Lee De Forest, the inventor of the triode electron tube – a mainstay in the development of modern radio. With an aerial and his new receiver, he observed from the roof of a building in Seattle, Washington, and surmised – probably erroneously – that the comet was the source of the enhanced crackling static he thought he had detected. In our time, radio observatories have provided evidence for cometary HCN (hydrogen cyanide) and CH_3CN (acetonitrile), organic cyanides that may be parent molecules of the cyanogen that frightened millions in 1910. The molecules and molecular fragments that have been found in cometary spectra are shown in Figure 8.12.

In the 1970s, radio spectroscopy opened a surprising new field when a wide variety of exotic molecules were found in interstellar space. You might, for example, be looking at some distant source of radio emission – the centre of the Galaxy, say – and discover a range of new spectral lines produced by the

The composition of molecules and molecular fragments identified in comets **8.12**, *(this page) and in the interstellar gas,* **8.13** *(next page). A key to the molecules identified is given opposite. There is a considerable similarity between cometary and interstellar molecules, but the molecules thus far identified in the interstellar gas tend to be much more complicated. (Diagrams by Jon Lomberg/BPS.)*

diffuse gas between the radio source and your radio spectrometer. The gas between the stars is very thin, but if you can add up all the molecules in a line of sight thousands of light years long, you might find even fairly rare molecules. This bonanza of interstellar molecules, most of them organic, is shown in Figure 8.13.

Astronomers tend to be nervous about the word organic – concerned that it might be misunderstood as a token of life on another world. But the word organic only refers to molecules based on carbon, and organic chemicals would be produced and destroyed in the universe even if there was no life anywhere. Circumlocutions such as 'carbonaceous' are in common use, but we will here employ the correct chemical term organic. While organic chemistry does not in any sense imply biology, nevertheless, if there are complex organic molecules produced somewhere in the universe, it may well have some bearing on the question of life elsewhere, or even life down here on Earth. Indeed, the origin of life on Earth probably involved pre-existing organic molecules that, by definition, could not have been made by living things.

There is, in the gas between the stars, an enormous diversity of organic molecules, some fairly complex. Ices and silicates are also known spectroscopically to be an important component of the tiny grains of interstellar dust. Since comets mainly live far from the planets, in the interstellar realm, it is not unreasonable that they should be made of interstellar stuff. If we assume that comets are composed simply of silicates and ices – frozen H_2O, CH_4, NH_3, CO_2, and so on – and allow molecular fragments to come off under solar radiation, calculations show insufficient C_3 and CN produced relative to other molecules. Silicates and ices alone cannot explain all the organic fragments discovered by William Huggins. But if it is assumed that the comets ultimately derive from interstellar matter of the sort shown in Figure 8.13, then the molecules detected in the comas and tails of comets can be understood.

Cometary silicates are probably intimately mixed with – perhaps coated by – complex organic compounds. The relative abundance of atoms in comets, so far as is known, is quite like that in the interstellar grains and gas. There is less carbon detected in the spectra of comets than there is – again in terms of relative abundances – in the spectra of interstellar grains. One possible explanation is that a large proportion of complex organic molecules in the cometary nucleus are involatile, or do not produce accessible spectra features. If this interpretation is correct, comets might be as much as ten percent organics. Only a few percent of dark organics would be sufficient to darken and redden the cometary snows sufficiently to accord with observation. But the best evidence for organic matter in comets is obtained from recovering small pieces of cometary matter, a subject treated in Chapter 13.

The picture which emerges from the work of Huggins and his successors at the spectroscope is this: a comet is a snowball covered with small mineral grains which are surrounded by complex organic matter. The organic molecules are distributed throughout the comet, although they may be concentrated in the surface layers. The amount of organics is at least a few percent – perhaps as much as ten percent – enough to dramatically darken and somewhat redden the snows. This chemistry is similar to what we know of the molecules between the stars, and is hard to understand unless the comet has been formed out of interstellar matter. Thus, more than any other body we have access to, comets are emissaries from the spaces between the stars.

Bar charts showing the relative abundances of key chemical elements in cometary ices 8.14 (left), in interstellar ices 8.15 (middle), and in the universe generally 8.16 (right). The atoms are hydrogen (light yellow); carbon (white); nitrogen (blue); oxygen (orange); sulphur (bright yellow); and silicon (brown). Hydrogen is greatly in excess in the universe compared to interstellar frost or comets. The comparison shows that cometary ices are very similar in atomic composition to interstellar frost. (Diagrams by Jon Lomberg/BPS.)

Chapter 9

TAILS

Hung by the heavens with black, yield day to night!
Comets, importing change of time and states,
Brandish your crystal tresses in the sky...

William Shakespeare,
Henry V, I,i.

This is a chapter about nothing, or something closer to nothing than anything in our everyday life. A cubic centimetre is the volume of a sugar cube. If you select a cubic centimetre of air in front of your nose and look extremely closely, you would find thirty billion billion molecules, all very small, vigorously running into one another. If, on the other hand, you were to make an inventory of the atoms or molecules in a cubic centimetre of a cometary tail, you would find a thousand or less – hardly any at all. A bright comet tail is much nearer to a perfect vacuum than the best laboratory vacuum our technology can produce on Earth. But even with very few molecules in a cubic centimetre of comet tail, there are a great many cubic centimetres. It is sometimes said that, like a genie in the Arabian Nights, a cometary tail stretching from world to world can be bottled up in a brass lamp. Newton wrote, '...a gaseous comet [i.e. the tail alone], with a radius a thousand millions of miles in length, if submitted to the same degree of condensation as the earth, could be easily kept in a good-sized thimble'. But, even if there was only one molecule per cubic centimetre, not a thousand, the enormous tail Newton imagined would require a thimble three kilometres across to contain it when compressed. 'Good-sized', indeed.

The tail is almost perfectly transparent: when it passes in front of a star you can always see the star through it. But still, it may seem astonishing that the air in the room should be transparent while the near vacuum in the comet tail is visible to the naked eye. The air is a something; we can feel it readily enough in a stiff breeze. The comet tail is very close to nothing. How can the something be invisible when the nothing is clearly seen? The key to answering this question is to remember that we see the comet against a black sky whereas the daylight sky is almost uniformly illuminated. Newton did a very difficult experiment, remembering the brightness of a

9.1 *(Opposite) A comet and its train wander among the other lights of the night sky. (From Grandville's* Un Autre Monde, *1844.)*

9.2 *Stars seen through the more distant reaches of the multiple-tailed de Cheseaux' Comet of 1744. (From Amédée Guillemin,* The Heavens, *Paris, 1868.)*

9.3 *Stars seen through the tail of the Comet of 1843.* (*From Amédée Guillemin,* Les Comètes, *Paris, 1875.*)

9.4 *The position of the tail of the Great Comet of 1843 as it sped around the Sun. P marks the position of perihelion. The tail (also shown in Figure 9.3) always points away from the Sun. (From Camille Flammarion,* Astronomie Populaire, *Paris, 1880.)*

comet and comparing it with the brightness seen in a laboratory experiment:

> Nor is the brightness of the tails of most comets ordinarily greater than that of our air, an inch or two in thickness, reflecting in a darkened room the light of the sunbeams let in by a hole of the window shutter.

A great many widely separated particles reflect light as well as the same number of particles compressed under the influence of the air. Still, what is it that we see when we look at the tail of a comet?

If you are fortunate enough to be standing under a clear sky with a comet readily visible above you, note the orientation of the comet's tail with respect to the Sun. If it is just before sunrise, note how the tail streams away from the glimmer of dawn on the eastern horizon. If it is just after sunset, the tail points away from the Sun, now located just beneath the western horizon. And if you are fortunate enough to witness a great daylight comet, it will then be entirely obvious that the comet tail points away from the Sun. There are no significant exceptions to this rule. Although from comet to comet there is some variation in the exact angle the comet's tail makes with a straight line from comet to Sun, this rule is, in general, as invariable as the observation that the horns of the Moon's crescent always point away from the Sun.

Comet tails point away from the Sun whether the comet is approaching the Sun or receding from it, a conclusion first drawn by Chinese astronomers during the apparition of Halley's Comet in the year 837. After perihelion passage the comet flies tail first out of the solar system. The tails of comets are much more like the effluvia from an industrial smokestack blown back on a windy day than the long hair of a bicyclist flowing behind her as she coasts down a hill on a windless day: it is not the motion of the comet through some resisting gas that determines the orientation of the tail, but something like a wind blowing out from the Sun.

There are two kinds of cometary tails: long, straight, faintly blue tails, pointing almost perfectly straight back from the Sun; and usually shorter, curved, faintly yellow tails. Before their nature was understood, they were called Type I and Type II tails, respectively (Figure 9.5). These designations are still in use. The Type II tails are yellow because they reflect

9.5 *The thin, straight Type I tails and the curved Type II tails of Donati's Comet over Paris, 5 October 1858. (From Amédée Guillemin, Les Comètes, Paris, 1875.)*

9.6 *Photographs of Halley's Comet located in the actual orbital positions and proper orientation. (Montage by Jon Lomberg.)*

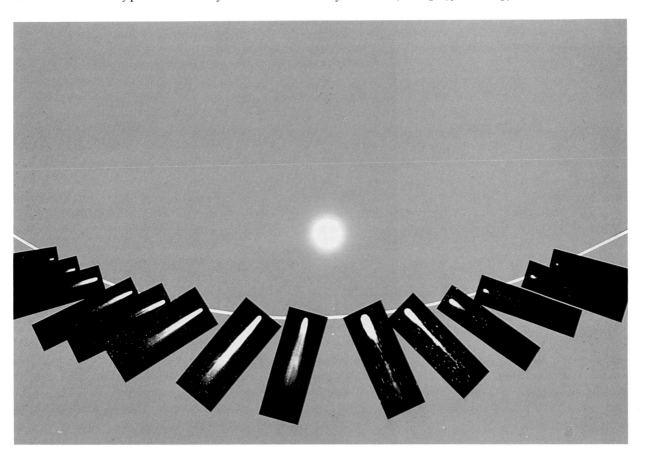

The tails of comets appear to be composed of the most volatile molecules which the heat of the Sun raises from their surface and by the impulsion of his rays banishes to an indefinite distance... The different volatility, size, and density of the molecules must needs produce considerable differences in the curves which they describe: hence arise the great varieties of form, length, and breadth observed in the tails of comets. If we suppose these effects combine with others which may result from a movement of rotation in the comet... we may partly conceive the reason of the singular phenomena represented by the nebulosities in tails of comets.

Pierre Simon, Marquis de Laplace,
The System of the World
(Paris, 1799).

THE COMET OF JULY 1819
to Mrs Bietta

*Oh what a fix! What a damned fix
I got into, if I'd only known, my husband and me!
I'm two weeks late with my...
So that... you know, Mrs Bietta!*

*And you know what? All this trouble on account of the Comet
With that tail that amazed everyone so!
The same thing also happened to Amelli,
And to Gina, and to Bina, and to Babetta.*

*And Nunziada's a full month late,
You know what a little fool she is:
Just imagine, that sneaky Comet!*

*And do you remember? We were there
laughing at that enormous tail
and playing under it! Can it be?*

Translated from the Italian
by Gina Psaki.

sunlight back to us; but the Type I tails give off a blue light of their own, although they are commonly not very prominent in visible light. A comet may have either type, or neither or both types of tails at a given time. The Type I tails often display an intricate dancing pattern of straight ray streamers, each ray narrower than the diameter of the Moon, but perhaps ten million kilometres long.

Type I tails are variable – not only from comet to comet, but from hour to hour, day to day, week to week, within the same comet. Like snakes and salamanders, a comet can grow a new tail: on the surrounding pages are examples, in drawings and photographs, of this propensity for regeneration. Comets are quick-change artists. Some of these pictures are printed as negatives, because black on white tends to bring out fine detail. In 1908, for example, Comet Morehouse – sizeable fragments of which were reported to have left the nucleus for the tail, where they emitted their own subsidiary tails – astonished the astronomers gathered at a conference in Oxford:

The formation of the tail seems to be intermittent rather than continuous. There seem to be at intervals convulsions or explosions in the nucleus, producing big bunches or lumps of tail which travel away and leave the comet with a small tail for a time... the signs of a coming convulsion can be recognized.

A young British astronomer named Arthur Eddington used the newly perfected lantern slide in a talk on Comet Morehouse:

Here is the same comet a day later. Everything is altered completely; you cannot point to a feature in the tail on this photograph and say that it corresponds to a certain feature in the previous photograph, and that one has changed into the other; you cannot say that this is the tail of the previous day modified. As far as can be judged the tail is an entirely new one...

Type I tails sometimes exhibit 'knots' – small condensations of matter brighter than their surroundings. Knots are sometimes observed to accelerate down the tail away from the Sun. In talking about Comet Morehouse, Eddington was clearly puzzled by the acceleration turning on and off capriciously. By taking a quick succession of photographs you can measure how fast the knots move. Their speed ranges up to 250 kilometres per second (540,000 miles an hour), and their acceleration (which you could feel if you were moving with the knot) can be as high as 1g, the acceleration that the Earth's gravity imparts to falling objects. There can also be knots with very little speed or acceleration. The motion of knots in comet tails is as unpredictable as the weather.

Type II tails are clearly made of innumerable fine particles. If only Newtonian gravitation were working, a collection of fine particles could not travel through space as if it were a solid body. Instead, each particle would be on a separate orbit about the Sun, a microplanet moving in an almost perfect vacuum. Because the initial speeds of the dust grains leaving the comet depend on the properties of the puff of gas that carried them off, some will be travelling a little faster and some a little slower than the comet. In their orbits, Mars travels more slowly than the Earth; and the Earth than Venus. In the same way, particles travelling faster will establish orbits closer to the Sun, and those travelling more slowly will be farther from the Sun. This slow gradient in velocity accounts, both qualitatively and quantitatively, for the characteristic curvature of the Type II comet tail. The great yellow tails of comets imply, by their very shapes, individual small particles in separate orbits around the Sun.

Spectroscopy of the Type II tail shows light from the Sun reflected back to the observer without the tail adding or subtracting spectral features of its own. This is characteristic of ordinary silicate dust, and in the spectrum of comet tails there is the infrared signature of silicates, the main constituent of ordinary rocks on Earth. For this reason, the Type II tail is called a dust tail, even though we suspect there is an intimate association of sticky, dark organic matter with the fine silicate dust particles.

Tiny particles – clusters of still smaller particles made of silicates and organics – are jetted off the cometary nucleus, and redirected back away from the Sun. Newtonian gravity will then produce the graceful, curved tail. Fair enough. But what is this mysterious influence driving the particles back? The first person to guess the right answer was Johannes Kepler, who held that the tail of a comet is pushed away by the pressure of sunlight. The net result, he argued, was that the comet would ultimately dissipate into interplanetary gas.

Radiation pressure is obviously not a factor in everyday life. Even very small people are not thrown to the ground by sunlight as they step out of doors on a cloudless day. The force of radiation pressure is the equivalent of the weight of a layer one atom thick at the surface of the Earth. Radiation pressure amounts very nearly to nothing. But if you are made of almost nothing in the first place, radiation pressure can push you around. The total force of the sunlight on us depends on how much area we present, but the resistance we offer to sunlight in its effort to push us around depends on our mass. The smaller a particle is, the more area it presents for its mass. Eventually, in free space, a sufficiently small particle will feel a stronger force of radiation pressure driving it outward from

9.7 *The tail of Halley's Comet, with enormously exaggerated brightness and extent, hanging over the city of San Francisco in April 1910. (Courtesy, Jerrod Metz.)*

9.8 *Forms of comets, drawn according to Pliny's written description. (From Amédée Guillemin,* The World of Comets, *Paris, 1877.)*

9.9 *The spectacular appearance of Comet Morehouse 1908 II as photographed by Max Wolf at the Heidelberg Observatory. (From N.B. Richter, The Nature of Comets, Methuen, 1963.)*

9.10 *A schematic representation of the pressure of sunlight blowing back the gasses and fine particles in the cometary coma to produce the tail. (Diagram by Jon Lomberg/BPS.)*

the Sun than the Sun's gravity pulling it inward towards the Sun.

Both the radiation pressure force and the gravitational force vary inversely as the square of the distance from the Sun; so once a particle is small enough for radiation pressure to dominate, it is continuously accelerated out of the solar system. If radiation pressure wins over gravity at the orbit of Mercury, it is still winning over gravity by the time the particle is half-way to the nearest star. But dust particles must be very small – about a hundred-thousandth of a centimetre across, too small to see even with an ordinary microscope – for radiation pressure to eject them from the inner solar system. Therefore, of the dust particles that are blown off the nucleus of a comet by the jets, only the smallest particles are driven back into space by sunlight. The bigger particles – having achieved escape velocity from the comet, but not from the Sun – establish individual orbits about the Sun.

When the particles are much smaller than this, they are even smaller than the wavelength of most of the sunlight, and the particles slip through the crests and troughs of the lightwaves and again are not driven outward. Thus, there is only a small range of particles, with sizes in the vicinity of the wavelength of yellow light, that are accelerated by sunlight out of the solar system. Interplanetary space should be depleted in particles of this size.

If the Type II tail is made of dust, what is the composition of the longer, still more prominent Type I tail? Aristotle had thought that comet tails bore some similarities to the aurora borealis, a view echoed by Kant:

> The Earth has in it something which may be compared with the expansion of the vapours of comets and their tails, namely, the Northern Lights, or Aurora Borealis... The same force of the rays of the Sun, which makes the Aurora Borealis, would produce a vapour head with a tail, if the finest and the most volatile particles were to be found as abundantly on the Earth as on the comets.

While it was not understood in the times of Aristotle or Kant, today we do know something about the aurora – the multicoloured time-variable pattern of lights in the sky, sometimes resembling rustling drapery – that is seen chiefly in the arctic and antarctic regions of the Earth. The aurora appears in the polar regions because it is there that charged particles – especially protons from the Sun – are guided by the

9.11 *A photograph of Comet Kohoutek taken on 20 January 1974. In this negative, the background stars are shown as points, not trails, because Comet Kohoutek was bright enough not to require a long time exposure. (Courtesy, NASA.)*

9.12 *The aurora borealis flickering in the Alaskan night sky. (Courtesy, NASA.)*

9.13 *Comet Morehouse 1908 II photographed on 25 November 1908 at the Royal Greenwich Observatory. (Courtesy, NASA.)*

9.14 *Turbulence in the tail of Comet Brooks 1893 IV. (Lick Observatory.)*

Earth's magnetic field. The protons pour into the atmosphere above the poles, tear the molecules of air apart a little bit, and induce them to glow. The aurora is variable because the supply of protons is variable. While the purported similarity in appearance between the aurora and the Type I tails of comets may be less than obvious, it is certainly true that electricity and magnetism are central to the understanding of both.

The spectra of comas mainly show, as we have said, such molecular fragments as C_2, C_3, and CN. But when the telescope points away from the coma and towards the tail, a completely different spectrum is seen, in which the lines of CO^+ are dominant. CO^+ is a molecule of carbon monoxide with an electron removed. Such electrically charged molecules are called 'ions'. Like C_2, CO^+ absorbs invisible ultraviolet light from the Sun which, through fluorescence, is re-radiated in blue light. This is why Type I tails are called ion tails, and this is why they glow blue. If you could somehow turn off the ultraviolet radiation of the Sun, the dust tail would look the same, but the ion tail would disappear altogether. There is only a relatively small amount of CO^+ in the tails of comets, but if you took the CO^+ away, the ion tails would also disappear.

Molecules are ordinarily electrically neutral; the number of negatively charged electrons on the outside of the atom just balances the number of positively charged protons in the inside of the atom. Consider a molecule of water sitting in interplanetary space near the Earth: it is struck by a photon of ultraviolet light from the Sun (or a proton), and one of its electrons is carried off into space. The molecule is therefore positively charged: it has more protons than electrons. The charged molecule is called an ion, and the process that carried off its electron is called ionization. If one electron had been stripped off, we would symbolize such a water molecule as H_2O^+, the plus sign indicating an excess positive electrical charge, or, equivalently, a deficit of one electron. If two electrons had been carried off, we could write this as H_2O^{++}. The same is true for CH^+, or N_2^+, or CO^+.

So molecule of CO, let's say, scurries off the cometary nucleus into space and is mugged by a light beam, losing an electron in the scuffle. Why should it now move straight back from the Sun? Why should the knots in the ion tails accelerate erratically? What emanation from the Sun does an ion see that a dust grain doesn't?

Radiation pressure is far too feeble to explain the acceleration of knots in the ion tail. Also, the acceleration of the knots varies with time, while the light put out by the Sun is extremely steady. There must be some other means, besides the light it radiates to space, by which the Sun can affect the tails of comets. If we study the knots over a period of months, we sometimes find a periodicity in the accelerations closely equal to the period of rotation of the Sun as seen from the comet. The conclusion is straightforward: the influence emanates from a particular region on the Sun, not from the Sun as a whole. The influence, whatever it is, reaches out to the comet only when the active solar region is oriented towards the comet. As the Sun rotates (once every twenty-seven days, as seen from the Earth), the active region on the Sun eventually faces away from the comet, and the ion tail acceleration subsides, only to revive as the region rotates around again.

It has long been known that the Sun episodically emits charged particles: a flare is seen through the telescope on the surface of the Sun, and three days later a 'magnetic storm' strikes the Earth, interfering with long-distance radio communications. Fortuitously, on 29 March 1943, the ion tail of Comet Whipple-Fedtke-Tevzadze exhibited a marked knot acceleration, and on the same day a major magnetic storm reached the Earth. From such observations, the German astrophysicist, Ludwig Biermann, working in Germany during and after the Second World War, calculated the properties of these bursts of charged particles from the Sun. His conclusion was that there must be a steady solar wind that blows outwards from the Sun at all times, and which drives the ion tails of comets leeward no matter what section of the sky the comet happens to be in. But in addition, the episodic accelerations in the ion tails must be due to solar flares that generate fluxes of charged particles much more intense than the ordinary solar wind, but which spread out in space only in certain directions.

The solar wind was first directly detected by the Soviet Luna 3 spacecraft in 1959, and later by the American spacecraft Explorer 10 and, especially, the interplanetary spacecraft Mariner 2, which made extensive measurements of the solar wind on its historic first voyage from Earth to Venus. The measurements showed that the solar wind consisted mainly of protons and electrons. There were only a few of them per cubic centimetre in the vicinity of the Earth, but they were rushing radially outward from the Sun at speeds of several hundred kilometres per second. In addition, occasional flares and other violent events on the Sun's surface propel globs of high energy protons and electrons out into space. This wind from the Sun carries with it a magnetic field which sweeps up ions in its path. The moment a molecule is ionized by sunlight, it is captured by the magnetic fields carried with the solar wind. However, those molecules that are not ionized, that remain electrically neutral, are unaffected by magnetic fields.

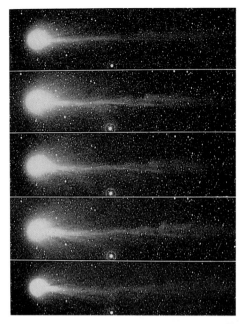

9.15 *Five photographs, taken approximately an hour apart, of Comet Whipple-Fedtke-Tevzadze 1943 I on 8/9 March 1943. Photographed in wartime Nazi Germany. (From N.B. Richter,* The Nature of Comets, *Methuen, 1963.)*

9.16 *A Moet & Chandon advertisement from 1910. (Courtesy, Library of Congress.)*

*...and the cork goes ceilingward,
the flow of comet wine spurts forth.*
Eugene Onegin, *Alexander Pushkin*

9.17 *Comet Mrkos photographed on 23 August 1957 from Mount Palomar Observatory. Note the structureless, faint dust tail and the extremely complex morphology of the ion tail.*

9.18 *A tail disconnection event in Comet Morehouse 1908 II. These three pictures were taken on three consecutive days – 30 September, 1 October, and 2 October – in 1908. A major piece of the tail has been lost downwind of the comet. Dashed lines are background stars trailed in this time exposure.*

Thus, the ion tails of comets are weather vanes in the solar wind, always blowing straight back; but, occasionally, an irregular magnetized cloud from the Sun catches up with the tail and stirs up the ions.

The ion tail has more variability, and, often, more structure than the dust tail. In a period of hours the thin, straight rays can change, merge, and dissipate. Evanescent features take complex forms, including sharp, right angle discontinuities, and corkscrew shapes millions of kilometres across. Some of the tail patterns (see Figure 9.19, for example) look almost as if the comet is veering erratically, like smoke bombs on the Fourth of July. But the comet is moving on an almost perfect elliptical orbit; the solar wind is responsible for the patterns. Sometimes, the entire tail disconnects itself from the nucleus and, fading, drifts slowly back, until it disappears entirely. Typically, the nucleus forms a new tail after a disconnection event, and the old and the new tail may interact, or even become briefly intertwined. The complexity and variability of the tails of comets has an almost biological feel to it. The prevailing opinion among astronomers is that the interaction of the electrically charged molecules in the ion tail with the highly variable solar wind accounts for most of the strange behaviour of comet tails. The comet, of course, contributes some idiosyncrasies of its own – as, for example, the timing and magnitude of gas and dust outbursts. But, in detail, we are still far from understanding the changing forms in the ion tails of comets.

Astronomers spend long hours arranging photographs of the same cometary tail in their proper time sequence, and – with equipment ranging from a magnifying glass to a computer contrast enhancement system – examining every burp and wiggle, in the hope of uncovering the underlying physical mechanisms. Since the flow of charged and neutral gasses through a moving stream of protons carrying its own magnetic field is not something scientists often run into in their everyday lives, the subject is necessarily arcane. It involves elaborate, three-dimensional magnetohydro-dynamic calculations, and laboratory simulations in which an electrically charged plasma interacts with some solid object (for example, a ball of wax) intended to duplicate the cometary nucleus.

The motion of comets in the Sun's gravitational field involves essentially the same physics as a rock thrown up into the air and falling back to the ground. There is a ready parallel in everyday life. The evaporation of ice from a cometary nucleus is not very different from the evaporation of snow on a sunny day. Again, we easily understand the phenomenon because it is part of everyday life. But – beyond a distant relationship to the aurora borealis – the plasma physics of cometary tails does not have a ready terrestrial analogue, so naturally it presents to us facets that are mysterious, that tend to resist our attempts to understand.

Our knowledge of the ion tails of comets is new and, in certain respects, still tentative, but it fully satisfies the expectations of astronomers of an earlier epoch. The American astronomer E.E. Barnard wrote in 1909:

> In any attempt to explain certain of these features of the comet, we are forced to tread on very dangerous ground, for there is nothing in the adopted theories of comets that will explain them, and we must assume unknown quantities that perhaps violate all our ideas of the conditions existing in space in the vicinity of the Sun and planets. But, as there seems to be no other explanation, the very fact that we are driven to extremes in the search for a possible cause may lead to a knowledge of interplanetary conditions that would never become known without the aid of the wide sweeping tail of a comet.

9.19 *Comet Morehouse 1908 II, with a view two weeks later than the tail disconnection event shown opposite, below. (Indiana University photograph, courtesy, NASA.)*

Today, we understand that the ion tails are solar windsocks, probes of interplanetary weather conditions which might otherwise escape our notice altogether.

After a comet begins to release gasses and a coma forms, the solar wind and the magnetic field it carries collide with the cometary atmosphere. Molecules in the outer layers of the coma are ionized by ultraviolet light from the Sun. The solar wind sweeps these ions up, flows around the sunward segment of the coma, and carries the ions far downwind. A shockwave is created, similar to that in front of a high performance aircraft when it reaches supersonic speeds. This bow shock is established as much as a million kilometres windward of the comet. The ions whipped leeward by the solar wind may stretch out a hundred million kilometres away from the Sun, pumped to glow an eerie blue by solar ultraviolet light. Occasionally, the tail is overtaken by a gust and more rarely by a small cyclone in the solar wind, and havoc is wrought in a tail that formerly stretched perfectly radially out from the Sun.

Because of their rapid and erratic variability, and technical and organizational limitations, there has never been an adequate time lapse motion picture of a comet tail. But now we have reached the point where this is possible, both from the ground and from a spacecraft. Even colour stereoscopic movies are now possible. Moreover, there are today a number of monitoring stations in interplanetary space that are routinely surveying the solar wind and its variations. We are now nearly ready to employ our knowledge of interplanetary weather in a comprehensive understanding of the long and graceful tails of comets.

9.20 *Comet Humason 1962 VIII photographed from Mount Palomar Observatory. Note the ray structure. (Courtesy, NASA.)*

THE COMET

March and Two Step

By

WILLIAM LEANDER SHEETZ

Piano .... .50
Band 1.00
Orchestra ... 1.00

PUBLISHED BY
THE WHITE-SHEETZ Cº
BURLINGTON, IOWA

Chapter 10

A COMETARY BESTIARY

If a rare [comet] and one of unusual shape appears, everyone wants to know what it is and, ignoring the other celestial phenomena, asks about the newcomer, uncertain whether he ought to admire or fear it. For there is no lack of people who create terror and predict dire meanings.

Seneca,
Natural Questions
Book 7, 'Comets'.

The astronomers... give more attention to the laws of their movements than to the strangeness of their form.

Immanuel Kant on comets,
General Natural History and
Theory of the Heavens (1755).

Like a basking whale before it plunges into the ocean depths, a comet briefly luxuriates in the sunlight, and then dwindles and is gone. The comets dance in the night sky, capriciously altering their brightness and their path.

Sometimes we make out extraordinary activity near the nucleus. Halos are produced, shed, and dissipated. Multiple fountains can be seen spraying gas and dust into space. Because the comet typically rotates once every few hours, the curvature of the streamers is marked. Everything is blown back from the Sun by radiation pressure and the solar wind.

Some comets propel a clockwork succession of shroud-like comas into space. Comet Donati is a famous example. The most ready explanation is that there is a patch of surface ice which explosively evaporates in sunlight; since the comet is rotating, a typical patch sees both night and day, and new comas are generated on the sunlit side just as often as the nucleus rotates.

In these pages we have accumulated a kind of cometary bestiary, like the animal bestiaries assembled by medieval authors to amaze and delight and even to instruct. Most of the animals displayed were real; many were exotic; a few, such as the unicorn, were errors in transmission, garbled accounts – in this case of the African rhinoceros. There were also some conscious hoaxes. We look back at the bestiaries with fondness, and perhaps a little amusement, recognizing that they are the forerunners of modern textbooks in zoology.

10.1 *(Opposite) A daylight comet portrayed on sheet music from 1909. (Courtesy, Ruth S. Freitag, Library of Congress.)*

10.2 *Successive envelopes of gas ejected from the nucleus of Comet Donati 1858 VI, drawn at the telescope on 5 October of that year by Schmidt. (Courtesy, NASA.)*

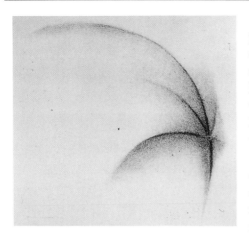

10.3 *Five (or, probably, six) jets off the nucleus of Comet Tebbutt 1861 II, drawn by Schmidt. (Courtesy, NASA.)*

10.4 *Schematic representation of an oblique view (left) and a view looking down on the axis of rotation (right) of the same jetting cometary nucleus. (Diagram by Jon Lomberg/BPS.)*

In a similar spirit, we offer here a range of images, photographed or drawn at the telescope, of the forms of comets; particular attention is paid here to the jets within the comas. Sometimes the comet presents itself so that we are looking down the axis of rotation; at other times, we are viewing obliquely. In all cases, it is easy to tell where the Sun lies. Some of the illustrations are reproduced black on white, to heighten contrast.

At the telescope, the astronomer is peering through an ocean of turbulent air, and the image produced is always distorted. One of the reasons that telescopes are located on isolated mountain tops, above much of the atmosphere, is to improve the seeing. At the focus of a large telescope, the astronomer's eye has a great advantage over the photographic plate: the astronomer can recall how, just a moment ago, the atmosphere steadied and fine detail in the comet could be glimpsed. By remembering such moments and drawing what was seen, the astronomer can often produce detail inaccessible to a camera attached to the same telescope. But the disadvantage of this method is that people are imperfect recording devices, and at the limit of resolution the eyes can play tricks. Nevertheless, by comparing the drawings of independent observers, by matching drawings with photographs, and by new image enhancement techniques, it is possible to verify that, by and large, the early cometary observers knew what they were doing.

Every one of these cometary forms represents a snapshot of the life and death of a comet, a glimpse of sunlight doing violence to a large lump of ice. Each of these displays is produced, at least in part, by streamers of gas and dust, rotation, radiation pressure, and the local weather conditions in the solar wind. The streamers and the coma are typically thousands of kilometres across, or more. Deducing the underlying events from the appearances left is what detectives do. It is also what field geologists do. Let your eye roam over the variety of cometary forms exhibited here, and see if you can construct a plausible explanation of what is happening in each case.

In Figure 10.3, we are looking obliquely at a nucleus while five or six streamers are being blown back by the Sun. For Figure 10.5 and elsewhere, the fountains are restricted to the heated sunward hemisphere of the cometary nucleus. The direction of rotation can often be deduced from the curvature of the streamers (Figure 10.5, for example).

In Figure 10.4 (left), we are looking obliquely at the multiple fountains of dust erupting from the day side of the Great Comet of 1861. But imagine that instead the comet had approached the Earth from behind, its pole by accident pointed in our direction. Then, we would be looking straight down on the axis of rotation, and the pinwheel of fountains would be rotating as we watched (see Figure 10.4, right).

Fortunately for us, the Earth's gravity isn't enough to drag passing comets to it. But every now and then, by chance, a comet will make a close pass by the Earth. It is only a matter of waiting. If you can wait long enough – a hundred million years, say – you are likely to see even very odd comets. For example, between the orbit of Saturn and Uranus is Chiron, a highly unusual object discovered in 1977 by Charles Kowal of the Hale Observatories in California. Named after the centaur who taught Zeus wisdom, Chiron is three or four hundred kilometres across, bigger than any known comet, although it is no larger than the bigger asteroids. Could Chiron be the most visible member of a previously unknown population of massive comets that live mainly beyond Pluto?

Chiron is very dark and red. Other objects in the outer solar system are darkened and reddened by complex organic matter, and it is likely that this is also the case with Chiron. It certainly cannot have much uncontaminated fresh ice on its surface. Perhaps Chiron is a comet whose surface ices of methane and other exotic volatiles have evaporated leaving a dark organic matrix behind.

Every few thousand years, Chiron makes a sufficient number of distant passes by Saturn for its orbit to be considered unstable. It may be slowly working its way in towards the Sun, and may one day become a short-period comet. Imagine, then, after repeated encounters with Saturn and Jupiter, that Chiron one day enters the inner solar system. If Comet Schwassmann-Wachmann 1 can still produce outbursts from its lonely post between the orbits of Jupiter and Saturn and, if the Great Comet of 1729 could be visible to the naked eye when it was almost at the distance of Jupiter, then what would an object like Chiron look like if it were newly arrived from the outermost solar system and making a close pass by the Earth? Doubtless there is water ice present that would explosively evaporate as Chiron approached the Sun. A very dark comet, perhaps just perceptibly red, hundreds of kilometres across, with multiple dust fountains and an immense tail, would be quite a spectacle as it passed by the Earth. There does not seem to be any historical record of such an apparition. It is doubtless a rare event.

Let us ask a more modest question: is there any historical record of the close approach to Earth of any sort of rotating, jetting cometary nucleus? In the case of Comet Morehouse 1908 III, the inner coma with its jets was about four thousand kilometres across, a more or less typical size, roughly that of the Moon. If such a cometary nucleus came as close to the

The thought suggests itself that the curves [in tail streamers] may be spiral curves produced by a rotation of the nucleus of the comet whilst it is discharging the streamers.

Arthur Stanley Eddington.

10.5 *Two views of the inner coma of Halley's Comet at different times of day on 5 May 1910, drawn in Johannesburg by R.T.A. Innes (left) and W.M. Worsell (right). The image on the right seems to imply rotation anti-clockwise to our line of sight. (Courtesy, NASA.)*

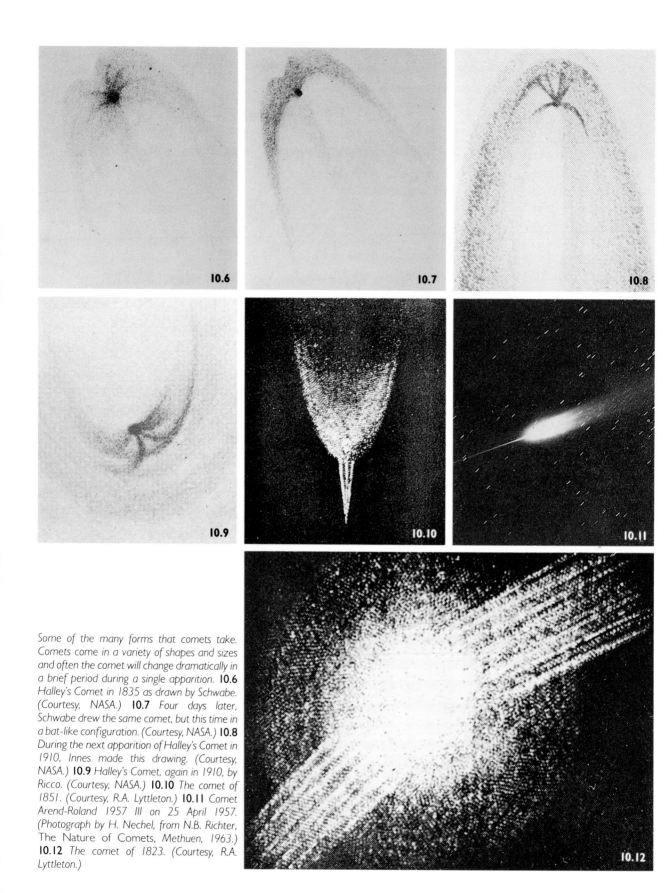

Some of the many forms that comets take. Comets come in a variety of shapes and sizes and often the comet will change dramatically in a brief period during a single apparition. **10.6** Halley's Comet in 1835 as drawn by Schwabe. (Courtesy, NASA.) **10.7** Four days later, Schwabe drew the same comet, but this time in a bat-like configuration. (Courtesy, NASA.) **10.8** During the next apparition of Halley's Comet in 1910, Innes made this drawing. (Courtesy, NASA.) **10.9** Halley's Comet, again in 1910, by Ricco. (Courtesy, NASA.) **10.10** The comet of 1851. (Courtesy, R.A. Lyttleton.) **10.11** Comet Arend-Roland 1957 III on 25 April 1957. (Photograph by H. Nechel, from N.B. Richter, The Nature of Comets, Methuen, 1963.) **10.12** The comet of 1823. (Courtesy, R.A. Lyttleton.)

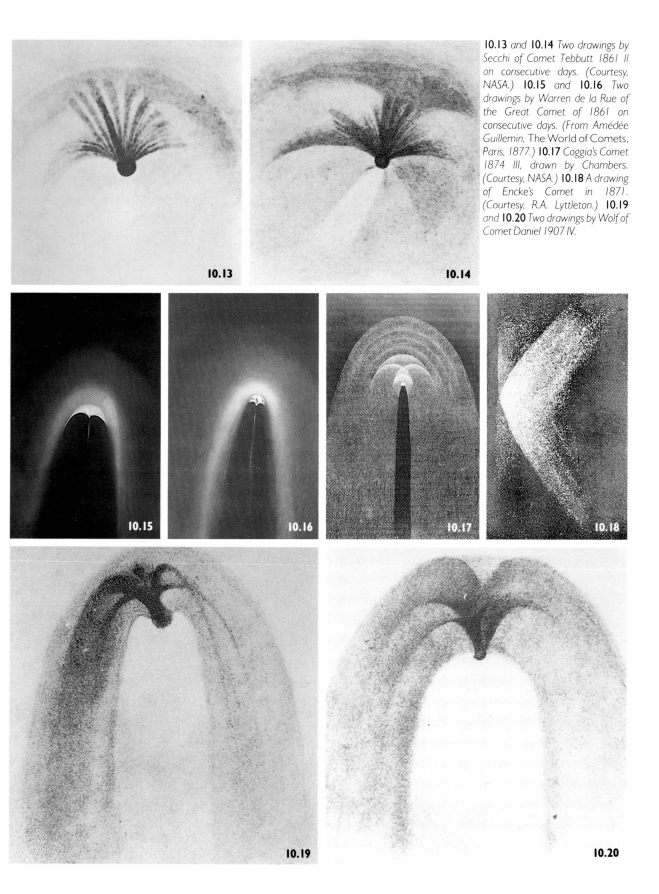

10.13 and **10.14** *Two drawings by Secchi of Comet Tebbutt 1861 II on consecutive days. (Courtesy, NASA.)* **10.15** and **10.16** *Two drawings by Warren de la Rue of the Great Comet of 1861 on consecutive days. (From Amédée Guillemin,* The World of Comets, *Paris, 1877.)* **10.17** *Coggia's Comet 1874 III, drawn by Chambers. (Courtesy, NASA.)* **10.18** *A drawing of Encke's Comet in 1871. (Courtesy, R.A. Lyttleton.)* **10.19** and **10.20** *Two drawings by Wolf of Comet Daniel 1907 IV.*

Earth as the Moon is, it would of course appear to us just as large as the Moon, about half a degree across. On 11 May 1983, Comet IRAS-Araki-Alcock (named after a spacecraft and two humans) passed within five million kilometres of the Earth, a dozen times the Moon's distance from the Earth. From the observed frequency with which new comets arrive in the inner solar system, it is possible to calculate how long we have to wait before a comet comes as close as the Moon. The answer is a few thousand years at most. If you're prepared to wait four or five thousand years, a cometary nucleus should pass you by at considerably closer range. Imagine the sky dominated for months by a dull, red, irregular object, spitting out white canopies, its shimmering, curved fountains flowing into space, and all the material eventually swept back into a vast tail that extends from horizon to horizon. It would be a memorable event.

Unless it made an approach to the Earth from an unlikely sector of the sky, the comet would be seen by cultures all over the world. Surely, there would be a mythological framework – sometimes called a world view – into which this apparition would be fitted? People would naturally think the display held some portent or significance for them. Some cometary form should, therefore, have entered the art of many cultures, perhaps even in a central way. As time passed, the memory of the true events might fade and the stories become vaguer, but the cometary form would still be a dominant motif in the art and records of the previous generations. If we saw such an apparition and believed it was a message for us, we would not be disposed to ignore it. After thousands of years, the cometary symbol, whatever it was, might be wholly disconnected from its bizarre and awesome origins. In a pre-scientific, pre-literate society, accounts of an unprecedented occurrence involving uncommon physics must necessarily, after thousands of years, take on lives of their own.

Because of extravagant claims made on meagre evidence in a succession of popular books and articles over the centuries, the notion that historical catastrophes have been caused by collisions of comets with the Earth has become disreputable. But, at least from Halley's time, it has been apparent that close approaches, and even impacts of comets with the Earth, will happen if only we wait long enough. As in all scientific questions, this one comes down to the quality of the evidence. We are aware of the need to proceed with caution.

The temptation to speculate on this matter is easily understood. If comets have come close by occasionally in the last few generations, then, over historical time, a really big comet must have been near, more spectacular than anything in living memory. So you scurry around in the art and literature of myth, or in the geological record, until you find something that seems attributable to a comet. Then you write a book. Thus, Ignatius Donnelly – Congressman, Lieutenant Governor of Minnesota, passionate advocate of human rights and the conservation of nature – proposed in 1883, in a wildly popular work called *Ragnarok,* that thick deposits of clay all over the world fell from the skies when the Earth passed through a comet built along the lines of the orbiting sand bank model (see Chapter 6). But clays are readily produced by known geological processes, and most show no sign in their atomic or molecular make-up of extra-terrestrial origin. Also, the sand bank model of comets is now known to be invalid, although it was the expert consensus in Donnelly's time. Or, as another example, a Russian-American psychiatrist named Immanuel Velikovsky proposed that the plagues of Egypt, manna from heaven, and other snippets from ancient myth were generated by a comet that came too close to the Earth. But a comet is not the only conceivable explanation of such stories, and other elements of his hypothesis are immersed deeply in error.

Yet, there are reasons to think that accurate depictions of extraordinary as well as ordinary cometary events were, at least on occasion, made by the ancients. For example, the historian Ephorus, in the fourth century BC, reported a comet that split into two. Subsequent writers continued to quote Ephorus, although with some degree of scepticism. By the time of Pliny, there were no other examples of fragmenting comets, and yet a record that a comet had once split in two was safely transmitted to our time. Today we know – as with Biela's Comet and the sungrazers – that fragmentation does occur. Ephorus is vindicated, and we learn that unusual cometary happenings can accurately be passed down through the millennia.

In our century, there have been extensive Newtonian calculations of the orbit of Halley's Comet, with predictions – or, rather, postdictions – of the date and position of the comet in the sky for every apparition back to 240 BC. For each one of these calculated apparitions, convincing evidence exists in ancient chronicles that a bright comet was indeed in the predicted part of the sky on the date in question. In addition to improving our already high confidence in Newtonian gravitational theory, this concordance between theory and observation renews our respect for the precision and attentiveness of ancient Chinese, Korean, Japanese, Babylonian and European chroniclers. While this respect does not extend so far as to giving full credence to descriptions of comets encumbered with dragons, swords or faces (see Figure

10.21 *Complex, almost biological forms near the jetting nucleus of Comet Donati, drawn at the telescope at Harvard College Observatory on 5 October 1858 by Bond. (Courtesy, NASA.)*

5, Chapter 2), it does suggest that in the historical depictions of the forms of comets there may be preserved for us something of the natural history of these idiosyncratic visitors from the depths of space.

Pliny noted the appearance of a comet 'too brilliant to be looked at directly; it was white with silver hair and resembled a god in human form'. What are we to make of this? For a comet to be so bright, it would have to be sheathed in a highly reflecting coma and pass close to Earth – not impossible. The configuration of the coma can be complicated, and can suggest a human form (for example, Bond's drawing of Comet Donati, Figure 10.21, which resembles a foetus). But a silver-haired comet god is certainly not widely dispersed in the myth and art of Earth.

Pliny described another kind of comet in these words: 'Like a horse's mane, it has a very rapid motion, like a circle revolving on itself.' The theme of rotation is occasionally connected with comets in the ancient records. Epigenes proposed that comets are born in whirlwinds, a view that Seneca dismisses for good reason, but a rotating, jetting cometary nucleus might look very much like a whirlwind. Among Seneca's objections are that whirlwinds have brief durations, and that their rotary motions would be dissipated by the speedy passage of the heavens around the Earth. Seneca did not know that the Earth turns. He says: 'The shape of a whirlwind is round... therefore the fire which is enclosed [within the hypothetical cometary whirlwind] ought to be like a whirlwind. And yet the first is elongated and scattered and not at all like something round-shaped.' It might just be that Seneca and Epigenes are describing different aspects of comets – Seneca the coma and tail, and Epigenes a close look at a rapidly rotating nucleus. (Before the invention of a telescope, the comet would have to come very close to the Earth for rotation to be apparent.) We therefore ask if there is some widespread ancient symbol, associated with the sky, that indicates rotation. Very tentatively, we suggest that there is one such symbol – the swastika.

This symbol of four bent arms emanating from a common centre was officially adopted and made synonymous with horror by the Nazi regime in Germany. The Nazi crimes against humanity are well-documented, if insufficiently reflected upon; long after their passing, the Nazis are still poisoning amity among nations. But there was a time, long before the Nazis, when swastikas abounded as benign symbols known to almost every culture on the planet. If you can, try to ignore the association of the Nazis with this symbol for a moment, and consider it for itself.

In 1979, for our 'Cosmos' television series, we were in India to film the ancient celebration of the cycle of the seasons, called Pongal. We were deeply moved by the generosity and kindness of the Hindu villagers of Thanjavur, in Dravidian, south India. We were a little taken aback, however, when they began cheerfully marking swastikas on their doorsteps in chalk. It was an ancient symbol of good fortune, they explained. Indeed it is.

In the two deepest and therefore oldest levels in Troy – dating back to 3000 BC in the Early Bronze Age – no signs of swastikas were found; but beginning with what he called the third, or burnt, city, dating back to the beginning of the second millennium BC, Heinrich Schliemann, the discoverer of Troy, found them everywhere. Hundreds of recovered artefacts, especially spindles operated in rotary motion, were festooned with swastikas. In Tang Dynasty China, public misuse of this important symbol had reached such a state that an imperial decree had to be issued forbidding the imprinting of swastikas on silk fabrics. In the west of India, there are caverns which serve as Buddhist shrines; most of the rock inscriptions are preceded or followed by a swastika. The Jains – a religious community in India which, in their respect for the sanctity of all life, stand at the opposite pole from the Nazis – use the swastika '…as a sign of benediction and blessing', for which reason also the Japanese once put it on their coffins. An account of Tibet in *The Times* of London in 1904 describes 'a few white, straitened hovels in tiers… On the door of each is a kicking swastika in white, and over it a rude daub of ball and crescent', astronomical symbols representing Sun and Moon. On blankets, beadwork, pottery, and other artefacts, the swastika was once a common, almost typical, symbol of the indigenous peoples of North America. Thomas Wilson, the Curator of the Department of Prehistoric Anthropology at the US National Museum, wrote in 1896:

> We know not whether it is intended as a religious symbol, a charm of blessing or good luck, or whether it is only an ornament. We do not know whether it has any hidden, mysterious, or symbolic meaning; but there it is, a prehistoric or Oriental Swastika in all its purity and simplicity, appearing in one of the mystic ceremonies of the Aborigines in the great American desert in the interior of the North American continent.

How did the same curious symbol become established in the ancient cultures of India, China, the American Southwest, Mayan Mexico, Brazil, Britain, and Turkey, among others? The swastika was in general use in Bronze Age Europe from the Arctic to the Mediterranean, spreading, in the Iron Age, to the Etruscan, Mycenaen, Trojan and Hittite civilizations. The word itself is Sanskrit:

10.22 *A spindle adorned with a swastika, one of hundreds found by Schliemann at Troy. (From Thomas Wilson,* The Swastika: The Earliest Known Symbol, and its Migrations; With Observations on the Migration of Certain Industries in Prehistoric Times, *Smithsonian Institution, Washington DC, 1896.)*

fountains in a cometary nucleus, recorded by generations of astronomers, to realize that there is here the potential for generating such a prodigy.

Imagine something like this: it is sometime early in the second millennium BC. Perhaps Hammurabi is King in Babylon, Sesostris III rules in Egypt, or Minos in Crete. More likely it is a time not today associated with any famous personage. While all the people on Earth are going about their daily business, a rapidly spinning comet with four active streamers appears. When people look up at the comet, they are looking down on the axis of rotation. The four jets, symmetrically placed around the equator on the daylit side, generate – because of the comet's rapid rotation – curved streamers, as you can easily see in the patterns formed by a rotary garden sprinkler. For the usual representation of the swastika, observers would have seen the pinwheel spinning anti-clockwise, with the arms trailing. As long as all four jets were on at once, people would see a brilliant swastika, perhaps somewhat foreshortened, in the daylight sky.

There is something anthropomorphic about the swastika. We read it as arms and legs in motion. It is one of the few symbols not ordinarily found in nature that is both simple and compelling. If something like it slowly materialized in your night sky amidst shrouds and fountains of dust – something self-propelled, animate, almost purposeful – you would surely find the experience noteworthy. You would speculate on its meaning, its religious significance, its portent. People would copy the symbol down so others would know about it, so that this marvel would not be forgotten. Whether you view it as an auspicious sign or as a harbinger of disaster, no one need explain to you that the thing is important.

That is how we imagine the swastika symbol might have become distributed worldwide. It's only a speculation, of course. The swastika form is not very different from the pinwheel structures seen in many comets, and brought out in

10.28 *Four symmetrically-placed jets in the illuminated (upper) hemisphere of a non-rotating cometary nucleus (left).* 10.29 *If, on the other hand, the nucleus is rapidly spinning (clockwise in this schematic representation), the arms trail the rotation and something like a swastika is produced. (Diagram by Jon Lomberg/BPS.)*

short exposure photography with a large telescope. Comet Bennett 1970 II is a recent example. For Comet Bennett, at least, the colour of the pinwheel was yellow, implying that the structure is in the dust. Looking at these forms, it will perhaps be granted that if enough spinning, jetting comets pass by the Earth, eventually one will present something like a swastika to view. But the argument alone is inconclusive. At least one piece of more direct evidence is needed.

Under these circumstances, it is arresting to find, in the culture with the longest tradition of careful observations of comets, a straightforward, apparently unambiguous description of a swastika as just another comet. Such is the case of the twenty-ninth and final comet to appear in the ancient silk atlas of cometary forms that was unearthed in a Han Dynasty tomb at Mawangdui, China (see Chapter 2). It dates from the third or fourth century BC, but is an anthology of observations that must be much older. The twenty-ninth comet is called 'Di-Xing', 'the long-tailed pheasant star'. The caption beneath the 'swastika' comet is the lengthiest of all twenty-nine, because the swastika comet alone is subject to differing interpretations. This comet is associated with change. 'Appearing in spring means good harvest, in summer means drought, in autumn means flood, in winter means small battles'. Of course, these auspices are fanciful, but they suggest more than one historical experience with comets of this form. So, the origin of the swastika in a cometary apparition seems a real possibility. Perhaps the connection is drawn in other little-noticed artefacts of ancient cultures.

The swastika achieved worldwide distribution, and was almost everywhere considered benign. Thousands of years later, the Nazi brand of racism and plunder arises and seeks a symbol to represent the allegedly superior, allegedly racially homogenous peoples of northern Europe. They already call themselves Aryans, after the white-skinned Persian invaders of dark-skinned India in the middle of the second millennium BC. The swastika is emblazoned on uniforms, weapons, stationery, aircraft, and assorted regalia. Children all over the world practise drawing the symbol. Under the banner of the swastika, the Nazis murder tens of millions and usher in an era in which humans are able to destroy their global civilization, and the human species as well. A pinwheel of evaporating ice appears in solitary magnificence in the skies of Earth and, 3,500 or 4,000 years later its form – remembered through all the intervening generations of humans – is employed in the cause of barbarism and murder. But there is an additional irony, because a cometary nucleus engaged in violent jetting is dying. Appropriately, the swastika symbolizes death.

Both the form of this luminous ejection, and the direction in which it issued from the nucleus, underwent singular and capricious alterations, the different phases succeeding each other with such rapidity that on no two successive nights were the appearances alike. At one time the emitted jet was single, and confined within narrow limits of divergence from the nucleus. At others it presented a fan-shaped, or swallow-tailed form, analogous to that of a gas flame issuing from a flattened orifice; while at others again two, three, or even more jets were darted forth in different directions.

John Herschel,
'On the 1835 Apparition of Halley's Comet' *Outlines of Astronomy* (London, 1858).

10.30 *The last seven comets shown in the Mawangdui silk from third- or fourth-century* BC *China. Nothing in the caption indicates the last form, in the shape of a swastika, to have been considered fundamentally different from the other comets shown.*

PART TWO

ORIGINS AND FATES OF THE COMETS

Chapter 11

AT THE HEART OF A TRILLION WORLDS

How many other bodies besides these comets move in secret, never rising before the eyes of men! For god has not made all things for man.

Seneca,
Natural Questions,
Book 7, 'Comets'.

We have reason to suspect that there are a great many more comets, which being at remote distances from the Sun, and being obscure and without a tail, may for that reason escape our observation.

Edmond Halley,
Transactions of the Royal Society of London, Volume 24,
Page 882 (1706).

The ancients imagined the planets to be attached to invisible machinery, wrought of transparent crystal spheres, elegantly coupled and geared. We now know that the ancients were wrong. There are no crystal spheres. The planets orbit the Sun guided only by the invisible hand of Newtonian gravitation. Some worlds are rock, some gas, some ice – and nowhere, from Mercury to Pluto, is there anything like a crystal sphere. But imagine ourselves leaving the solar system at some impossible speed, until even the orbits of the outermost planets are too small for us to see, until even the Sun is only a point of light no brighter than the brightest stars seen from Earth. Then we do encounter something like a crystal sphere, but a shattered one – a cloud of a trillion shards and fragments of ice, little worlds each the size of a city, feebly illuminated in the great dark between the stars.

We live at the heart of a trillion invisible worlds. It sounds like the teaching of some New Age sect. And we are not talking of metaphorical worlds, but rather of a trillion places, every one of them as distinct a world as ours is, every one gravitationally bound to the Sun, every one with a surface and an interior and even, on occasion, an atmosphere.

Cast your eye upwards. Concentrate on the smallest piece of sky you can make out. Imagine it extending in a widening wedge far out into space, to the stars. In that little patch of sky are a hundred thousand worlds or more, worlds unseen, unnamed, but in some sense known. These distant cousins of the Earth are the cometary nuclei – cold, silent, inactive,

It seems however not agreeable to the uniformity of the universe, that after a short view of the Sun [the comets] should be continually flying farther off, in that wide void beyond the planetary bounds, to creep along that dark cold region for millions of years... but that they should rather revolve about the Sun, in certain, though long periods.

Thomas Barker,
Of the Discoveries Concerning Comets
(London, Whiston and White, 1757).

11.1 *(Opposite) Schematic representation of the Oort Cloud of comets surrounding the Sun. Only a tiny fraction of the trillion comets in the outer Oort Cloud are shown, most of them lying about one-third of the way between the Sun and the nearest star. Much further in towards the Sun is the probably still more numerous cometary population of the inner Oort Cloud. (Painting by Jon Lomberg/BPS.)*

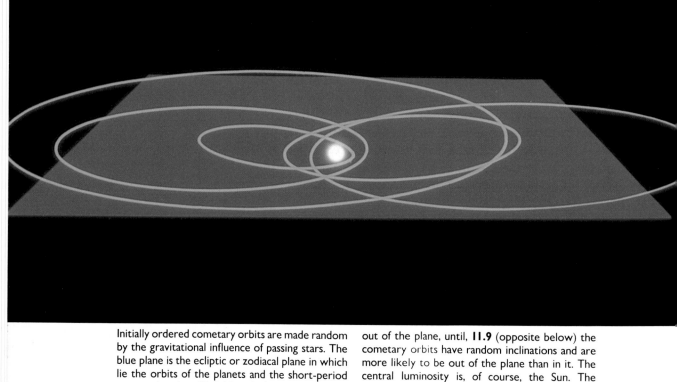

Initially ordered cometary orbits are made random by the gravitational influence of passing stars. The blue plane is the ecliptic or zodiacal plane in which lie the orbits of the planets and the short-period comets. Imagine, **11.6** (above) that comets in the Oort Cloud and long-period comets were originally restricted to this plane; as, through the age of the solar system, **11.7, 11.8** (below, opposite above), stars pass by, the cometary orbits are lifted out of the plane, until, **11.9** (opposite below) the cometary orbits have random inclinations and are more likely to be out of the plane than in it. The central luminosity is, of course, the Sun. The gravitational influence of these passing stars serves also to propel an occasional comet in the Oort Cloud into the inner solar system. (Diagrams by Jon Lomberg/BPS.)

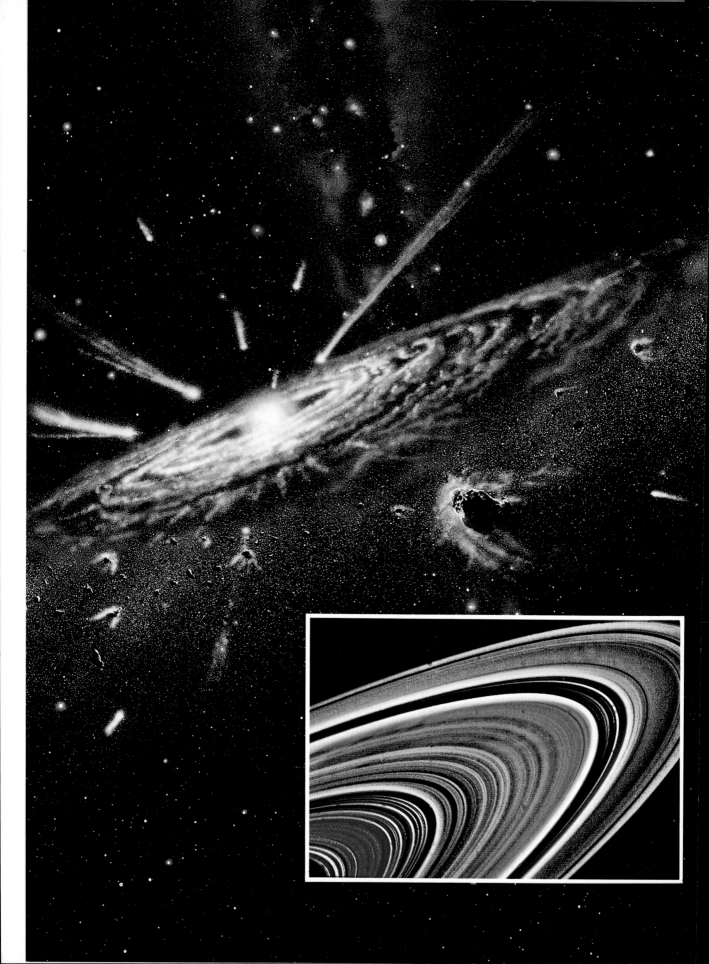

Chapter 12

MEMENTOS OF CREATION

Out of whose womb came the ice?
Job.

*It often happens that comets arise. These... are not any of the stars
that were made in the beginning, but are formed at the same time by
Divine command and again dissolved.*

John of Damascus,
*Exact Exposition of
the Orthodox Faith*
(eighth century).

Humans woke up one morning, looked around and discovered that our star, the Sun, has a retinue of other worlds, none just like the Earth. The inner, nearby, so-called terrestrial planets (Mercury, Venus, Earth and Mars) travel on nearly circular orbits in the same direction around the Sun as the Sun's own rotation.* Their silicate mantles conceal their metal cores.

Beyond them are the asteroids, thousands of small irregular objects – some rocky, some metallic, and some rich in dark, complex organic matter. The biggest are hundreds of kilometres across, but many more are a kilometre in size or smaller. They also move in prograde motion around the Sun, closely confined to the plane within which all the other planets perform their circumnavigations. Most of the asteroids are in nearly circular orbits between the paths of Mars and Jupiter. A few have more elliptical orbits that carry them inside the orbits of Mars or the Earth or Venus.

Farther out are the giant gas planets, Jupiter, Saturn, Uranus and Neptune, made mostly of hydrogen and helium, but with smaller amounts of water, ammonia, and methane. They vary in mass from fifteen to 300 times that of the Earth, and appear to have small interior cores of rock and metal. Orbiting these so-called 'jovian' planets are dozens of moons – each constructed from some different proportion of rock, ice, and organics. In this realm and far beyond are trillions of icy comets, most of them in circular orbits enormously distant from the planets; half are in prograde revolution, half in

12.3 *A terrestrial planet in retrograde motion round the sun. (Diagram by Jon Lomberg/BPS.)*

12.1 *(Opposite) A snapshot from the origin of the solar system almost five billion years ago. The flattened disc surrounding the young Sun is the solar nebula, in which condensation of enormous numbers of kilometre-sized objects of snow and rock are forming. Some of them (only a few are shown here) develop tails and can certainly be described as comets. It is from the solar nebula that all the planets, moons, asteroids, and comets in the solar system today are thought to derive. (Painting by Jon Lomberg.)* 12.2 *(inset) The rings of Saturn as observed by a Voyager spacecraft – Saturn is just out of the picture in the lower left-hand corner. The minor colour differences between these many flat rings have been greatly enhanced by computer processing. The large dark gap in the rings about three-quarters of the apparent distance to the periphery in this photograph is the Cassini Division. For Kant and Laplace, the rings of Saturn served as a model for the structure of the solar nebula. (Courtesy Nasa.)*

*This direction of rotary motion – clockwise if you were looking down on the solar system from high above the North Pole – is called direct or prograde. Motion in the opposite direction is called retrograde (see Figure 12.3).

retrograde, their orbits at random inclinations. A much smaller number of comets have highly eccentric orbits that take them near the inner terrestrial planets, where still fewer short-period comets can be found. They are all orbiting in a prograde direction and within the ecliptic plane. Now, where did this profusion of worlds come from? Why the regularities in their motions?

On first hearing where babies come from, many of us express a certain scepticism. Often, what is being described does not correspond to the listeners' own observations. Compared to storks, angels and cabbage patches, the machinery seems unwieldy and improbable. But no matter how implausible the story may seem at first, there is a reasonable consensus among knowledgeable individuals on the matter, and direct experience eventually converts even the most committed sceptic.

In the study of cometary or planetary origins, exotic processes over immense distances and in remote epochs are routinely invoked. If we are sceptical of an astronomical theory of our origins, we cannot poll the opinions of reliable astronomers on nearby worlds. There is no one to tell us about the cosmic version of the birds and the bees. We must figure it out for ourselves. Astronomers have a potential advantage, though: if they are curious about the stars, they can examine billions of them. Even a very unlikely process can be witnessed if enough stars are examined. Failing that, astronomers are reduced to the fundamentals – what is already known about the universe, and about the principles of physics.

Something like the modern view of the origin of the solar system was first proposed by two extraordinary thinkers whom we have already met in these pages – Immanuel Kant and Pierre Simon, the Marquis de Laplace. They had a distinct advantage over their predecessors: in the eighteenth century, observational astronomy was making remarkable discoveries. Kant and Laplace were intrigued by the structure of the rings of Saturn as had been revealed by Galileo, Huygens, and their successors. Here was a planet with a flat disc of fine particles surrounding it in its equatorial plane. Did the Sun once have a much larger ring system, from which the planets somehow condensed?

Kant had accepted and further developed Thomas Wright's insight that the Milky Way is a thin plate of stars, one of which is the Sun. Astronomers were discovering strange, flattened, luminous forms in the night sky – spiral nebulae they were later called. (Nebula, plural nebulae, is a Latin word meaning cloud.) Nature seemed to have a propensity for flattening extended systems, whether made of dust or of stars.

12.4 *(Opposite) Rings of other planets. Saturn has an elegant ring system composed of enormous numbers of fine particles orbiting the planet in its equatorial plane. The rings of Saturn have been known since the eighteenth century. But recently it has been discovered that other giant planets have equatorial ring systems. Above left and right are two false colour images of the rings of Jupiter, taken by the Voyager spacecraft and contrast-enhanced by Mark Showalter of Cornell University. Below is the earliest unambiguous image of the rings of the planet Uranus, taken in the infrared, where the rings are relatively bright and the planet is relatively dark; it was taken at the 200-inch Hale telescope at Palomar Observatory on 26 May 1983 by Philip Nicholson of Cornell University and Keith Matthews of the California Institute of Technology.*

However arbitrary the system of the planets may be, there exists between them some very remarkable relations, which may throw light on their origin; considering them with attention, we are astonished to see all the planets move around the Sun from west to east, and nearly in the same plane, all the satellites moving around their respective planets in the same direction, and nearly in the same plane with the planets. Lastly, the Sun, the planets, and those satellites in which a motion of rotation have been observed, turn on their own axis, in the same direction, and nearly in the same plane as their motion . . .

A phenomenon so extraordinary is not the effect of chance, it indicates an universal cause . . .

Laplace, 'On the
Regularity of the Solar
System', *System of the
World*, Part I, Chapter 6.

To this he added two more regularities, the near circularity of the orbits of the planets, and the great eccentricity and random inclinations of the orbits of the (long-period) comets.

The Kant-Laplace hypothesis for the origin of the solar system involves the interplay of rotation and gravitation. Imagine some irregular cloud of interstellar matter made of gas and dust and destined to form the solar system. All such clouds known today exhibit slow rotation. If the cloud is sufficiently massive, the random molecular motions are overwhelmed by self-gravity – the mutual attraction of the atoms and grains in the cloud to one another. The cloud begins to contract, the distant provinces falling inwards, the density of the cloud increases as a fixed amount of matter squeezes itself into progressively smaller volumes. As it contracts, the cloud spins faster, for the same reason that a pirouetting ice skater speeds up as she brings her arms in. (The experiment can also be done with a small person, seated on a rotating piano stool, holding a brick in each outstretched hand, and then rapidly drawing them in. This demonstration must be performed with caution.) The principle of physics involved is called the conservation of angular momentum, and is derived from Newton's laws of motion.

But as gas and dust and occasional condensations in the cloud spin faster, they experience an increasing reluctance to continue falling inwards, a phenomenon called centrifugal force (centrifugal meaning fleeing the centre). A pail of water on a string whirled sufficiently fast around your head does not spill – at least, not until you stop whirling. The centrifugal force balances gravity.

There is a type of amusement park ride in which a hollow cylinder is rapidly spinning, with a crowd of laughing, screaming people – poised somewhere between terror and delight – glued by centrifugal force to the rotating interior surface. When the cylinder stops spinning, the people come tumbling down off the walls.

The contracting cloud will also experience the centrifugal force, which will slow it down and eventually stop the contraction – but only in the plane of rotation. If you are standing on a small lump of matter falling towards the centre of the cloud, but along the axis of rotation (rather than in the equatorial plane) you do not feel any centrifugal force. The result is that matter in the equatorial plane stops collapsing, while matter along the axis continues to fall in. As a result, an initially irregular cloud becomes, in time, a flattened disc. The further the disc collapses, the more rapidly it rotates, and the denser it becomes at the very centre. The collapse stops, or at least slows, when the disc is spinning so fast that matter spews off at the periphery.

The Kant-Laplace hypothesis proposes that, long ago, an irregular, rotating interstellar cloud collapsed in this manner, with the central condensation forming the Sun. There is no doubt today that interstellar matter compressed to the density and temperature of the Sun will initiate thermonuclear reactions and begin to shine like a star. But it was a daring hypothesis for the eighteenth century. Other, smaller nearby condensations, Kant and Laplace proposed, formed the planets, each sweeping out a wide swathe of adjacent debris as it grew in size. The result would be a regular spacing of the newly-formed planets, something like the layout of the solar system today. Still smaller condensations near the planets formed their moons. The idea behind the Kant-Laplace hypothesis is more important than the precise details: they proposed that the solar system *evolved* – from a very different primordial state and without stellar, much less divine, intervention.

Because the word nebula means cloud, and out of analogy with the spiral nebulae (which are, of course, of much larger, galactic dimensions), the contracting cloud that formed the Sun and the planets is traditionally called the solar nebula. Today, we know a much larger variety of flat, rotating clouds around the nearer stars. It is more fashionable to call them accretion discs.

Laplace suggested that during the formation of the solar system, the Sun's atmosphere once extended far out into space, perhaps in consequence of an enormous explosion in the Sun, like the supernova of 1572 produced by a star in the constellation Cassiopeia. Or perhaps it was the residuum of the original solar nebula. Laplace's interstellar comets were, he imagined, falling in towards the Sun. The material in the solar nebula slowed comets down in the inner solar system, altered their orbits, and induced them to impact the Sun. The drag of the solar nebula cleaned the inner solar system of comets with nearly circular orbits, but left the comets at much greater distances unaffected. Through gravitational perturbations by the jovian planets, an occasional comet is induced to visit the inner solar system. The idea is remarkable in several respects. It indicates a kind of natural selection in the physical world well before Darwin put forward the idea; it proposes that there were once many more objects in the solar system than there are now; and it hypothesizes a large repository of comets beyond the most distant planet known.

Why, then, were the planets not similarly disturbed, and induced to collide with the Sun? Laplace proposed that the planets had formed from successive condensations in the early solar nebula. A ring of empty space, centred around the orbit of each new planet, was formed as the planet, growing at the expense of adjacent material, swept its surroundings clean of nebular debris. Perhaps he toyed with the idea that dark breaks should exist in the rings of Saturn if there are moons among the rings. However, he urged caution in accepting his hypothesis, which he offered 'with that distrust which every thing ought to inspire that is not the result of observation or calculation'. Probably because he was convinced of the interstellar origin of comets, it seems not to have occurred to him that comets, as well as planets, might condense out of the solar nebula.

That the rotation and revolution of the satellites are in the same direction as the rotation of their planets; that the planets rotate in the same sense that they revolve; and that the orbits of the planets are close to circular, while the comets have highly eccentric orbits – all followed naturally if everything (including or excluding the comets) had condensed out of the same rotating and collapsing cloud.

The hypothesis of Kant and Laplace is seen to be one of the happiest ideas in science, which at first astounds us, and then connects us in all directions with other discoveries...

'On the Origin of the Planetary System', Lecture delivered by H. Helmholtz, in Heidelberg and Cologne, 1871, *Popular Lectures on Scientific Subjects*, H. Helmholtz, (D. Appleton and Company, New York, 1881).

For both Kant and Laplace, the nebular hypothesis explained the regularities of the solar system as the end result of an evolution of worlds. Both believed that many other stars were surrounded by planetary systems which had evolved from their own accretion discs. In the last few years, ground-based and space-borne observations have revealed that many nearby stars are surrounded by accretion discs. The initial discovery was made by a space observatory called IRAS, the Infrared Astronomy Satellite, a joint Anglo-Dutch-American endeavour. Vega is one of the brightest stars in the night sky, only twenty-six light years distant, and it was a real surprise to discover that this well-studied star is surrounded by a disc of debris. It showed up as an elongated source of infrared radiation centred on Vega. Now, Vega is a star considerably younger than the Sun. To find an accretion disc around Vega strongly suggests that most, perhaps even all, ordinary stars are surrounded by such a disc during and immediately after their time of formation. Something eventually tidies up the disc – perhaps a combination of radiation pressure, the stellar wind, and planetary formation. But it takes time. And in that time, additional bodies may be condensing out of the nebula.

IRAS also provided infrared evidence of an accretion disc around a star called Beta Pictoris, among many. Soon afterwards, Bradford Smith, of the University of Arizona, and Richard Terrile, of the Jet Propulsion Laboratory, attached a special, highly sensitive camera, developed for Space Telegraph, to a ground-based telescope, and were able to photograph the Beta Pictoris accretion disc in ordinary visible light. The disc extends at least 400 Astronomical Units from the central star (here blocked out, so its radiation will not overwhelm and wash out the much more feeble light reflected off the disc). If this were a picture of the Sun in its early history, the accretion disc would extend much further from the Sun than does the orbit of the farthest known planet, some thirty to forty AU out. Smith and Terrile deduced a relative absence of debris in the interior of the disc, and suggested that this region had already been swept up by the condensation of planets – that are, however, much too small to be seen directly. Accretion discs have also been found around infant stars formed only a million years ago.

Thus, it now seems that the Kant-Laplace hypothesis is, in its fundamentals, verified – and by a technology that would have delighted both of them. The Sun, the planets and their moons all condensed out of the same rotating and collapsing disc of gas and dust. This is why all the planets revolve in the same plane in which the Sun rotates. Newton's view that the regularity of the planetary motions is direct evidence of Divine intervention has been superseded by another, more evolutionary view – still determined, to be sure, by the laws of nature, which, if we wish, we may attribute to a god or gods. But when queried – by Napoleon, of all people – about why his account of the origin and history of the solar system made no

12.7 *What may be another solar system in the late stages of formation is shown in this picture of the star Beta Pictoris, masked at the telescope at the centre of the picture in order to bring out the fainter disc of debris (yellow and pink diagonal feature). This faint disc, seen nearly edge-on, may be no more than a few hundred million years old. The picture was taken by Bradford A. Smith (University of Arizona) and Richard J. Terrile (Jet Propulsion Laboratory) at the 2.5-metre telescope of the Las Campanas Observatory, near La Serena, Chile.*

mention of God whatever, Laplace replied: 'Sire, I have no need of that hypothesis.'

Let us now follow a modern rendition of the Kant-Laplace hypothesis, in which we pay special attention to the origin and evolution of the comets. From direct spectroscopic evidence, we know the interstellar gas to be composed mainly of hydrogen and helium, although it is rich in many other materials, including complex organic molecules. Besides the gas, the other chief constituent of interstellar space is an enormous number of motes of dust. One of them, placed on a table before you, would be entirely invisible. They are, typically, a tenth of a micron across. But concentrate enormous numbers of them over hundreds or thousands of light years, and you can have enough dust to blot out the stars. The chemistry of the grains can also be inferred. Most seem to be made of ices, silicates, and organics – very roughly in equal proportions. Since this mixture of gas and grains makes up the interstellar clouds everywhere in the Milky Way, it must also have constituted the early collapsing solar nebula. Since interstellar space ordinarily holds much more gas than grains, this should have been true for the solar nebula as well.

As the nebula contracts and its density increases, collisions of grains with one another become more frequent. In part because of the organic content of these grains, when they collide they tend to stick. Big grains gobble up smaller ones. But all this does not go on in the dark. The primitive Sun has begun brightly shining. In the outer parts of the disc, it is still sufficiently cold for exotic ices such as methane or carbon monoxide to be perfectly stable in the growing condensations of matter. But in the very inner solar system, it is too hot even for water ice; the ices on the grains evaporate and dissipate, and what survives is mainly made of silicates. You have to carry a rock very close to the Sun for it to boil. As a result of all this, the chemistry in the inner solar system must have differed from the chemistry in the outer solar system – silicates predominating inside, and ices with a few percent of organics outside.

According to several calculations, a vast number of kilometre-sized objects should have accumulated throughout the nebula – silicate-rich ones on the inside, ice-rich ones on the outside. These objects should have been generated, not primarily by the grain-by-grain collisions, but by a fundamental instability in the solar nebula, in which objects a few kilometres in size were quickly and preferentially formed.

Both dust grains and gas gravitationally collapsed to form the disc. But it takes a great deal of gravity to hold on to so lightweight and thus fast-moving a molecule as hydrogen. In the middle part of the nebula, the kilometre-sized lumps

collided and grew into still larger objects, until a few aggregations of matter were able to retain the cold gas around them. This was the evolutionary line to the jovian planets. The original accretion core is smothered in a vast sphere of gas. In the warmer inner solar system, the grains, divested of their ice, grew more slowly and the temperatures were higher – both effects making it more difficult for the gas to be captured by the growing rocky spheres. This was the evolutionary line to the terrestrial planets.

Big objects would sweep up smaller ones on adjacent orbits. Because the relative velocities were low, the two bodies would tend to collide softly and merge. Eventually, a few large objects were produced, in orbits that never intersect. These became the planets. There is a kind of collisional natural selection at work here. You start out with a large number of growing objects in chaotic orbits, but through a process of collision and growth, and only occasionally the shattering of worlds, the solar system becomes regularized, simplified. The number of worlds steadily declined, from trillions to thousands to dozens. If you look at the planets today, you find them decorously spaced, their orbits by and large almost perfectly circular – except for the case of Pluto*, planets give

12.10 *Schematic illustration of condensation in the early solar nebula. We are looking down on the nebula, its bright, hot interior shown on the left. The farther from the forming Sun, the cooler it is. Three different materials are shown: methane (CH_4) in green; water (H_2O) in blue; and silicates (SiO_2) in orange. The cubes indicate solids, the clouds indicate vapours. In the outermost portions of the solar nebula, methane condenses as a solid; farther in, it is present only as the gas. Water is solid as ice well into the interior of the solar system, and silicates survive as solids almost to the surface of the Sun. Thus, rocky planets were formed in the interior of the solar system and icy bodies farther out. (Diagram by Jon Lomberg/BPS.)*

*And for this reason some scientists have supposed that Pluto is an escaped satellite from Neptune, returning periodically to the scene of the crime.

each other wide berth. Those early bodies on highly eccentric orbits were in danger; very soon, they collided with a world or were ejected from the solar system. Eventually, the only planets left were those that had, by chance, developed on orbits that quarantined them from their neighbours. It is just as well for us that they did; frequent world-shattering collisions are probably not good for the development of life.

The planets so formed would be orbiting the Sun in the sorts of orbits we recognize for the planets today. While no one has been able to prove that exactly nine planets should form – and not, say, six or forty-three, the entire question of the number of planets finally made being a matter of collisional statistics – the general picture is very successful, and explains not only the orbits, but also the overall chemical differences between the terrestrial and the jovian planets that we observe today.

If you picture the collapsing disc of gas and dust flattening and spinning faster, the grain collisions generating larger and larger objects, with the eventual formation of kilometre-sized objects which collide and grow still further, a question may occur to you: what happened to all those kilometre-sized objects? Are there any left? Were they all swept up as the growing planets ran into them, or might some of them still exist somewhere, unchanged from the epoch in which the solar system was formed?

When we calculate the fate of that original population of

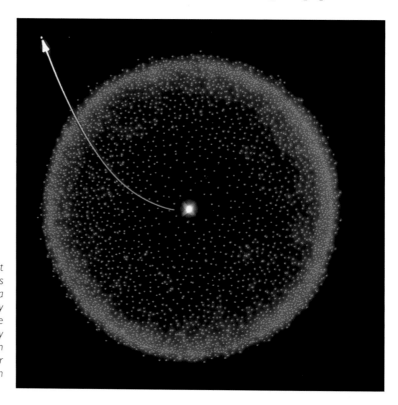

12.11 *In the early history of the solar system, the Oort Cloud is formed at least in part by ejection of comets from the vicinity of the giant planets. The solar nebula and the giant planets are much too small and far away in this picture to be seen. Cometary bodies in the vicinity of Uranus and Neptune are mainly gravitationally ejected to form the Oort Cloud, shown here in blue. Cometary bodies in the vicinity of Jupiter and Saturn are ejected from the solar system altogether (arrow). (Diagram by Jon Lomberg/BPS.)*

small worlds, we discover that gravitational interactions with the newly-finished jovian planets would have ejected multitudes of kilometre-sized worlds into the outermost gravitational frontier of the solar system, like an automatic pitching machine throwing baseballs into the stands once a minute for a hundred million years. This is how the Oort Cloud was generated. There is a population of primitive bodies that, four-and-a-half billion years ago, were sequestered so far from the Sun that no vaporization, no collisions, nothing at all could transform them. They are the stuff from which the solar system was formed and they are waiting for us. The Oort Cloud, and any newly arrived comet, is the answer to an astronomer's dream.

In the 1970s, V.S. Safronov, a Soviet specialist in the early history of the solar system, and J.A. Fernandez, a young Uruguayan astronomer, showed that if primitive cometary bodies (those kilometre-sized objects) were formed in the vicinity of Jupiter and Saturn, gravitational perturbations by these massive planets would eject them out of the solar system altogether. But if the parent bodies of the comets tended to be born in the vicinity of the less massive planets, Uranus and Neptune, their gravitational influence would, in general, eject the cometary bodies into the Oort Cloud, but not out of the solar system. So if these primitive worlds had condensed throughout the solar system, most would have been used up in making planets and in being ejected into interstellar space. But trillions, at least, would have been redirected to the Oort Cloud.

If the parent bodies of the comets had been formed in the vicinity of Jupiter, exotic ices would not have survived; if still closer to the Sun, even ordinary water ice would not be retained. Thus, two independent considerations – making the primitive comets out of the right stuff, and ejecting them into the right orbits – point to an origin in the rough vicinity of Uranus and Neptune.

Comets, it seems, were formed ultimately out of interstellar grains and gas within the solar nebula, just a little before the moons and planets formed, some 4.6 billion years ago. Many comets collided with each other, forming larger bodies, and sacrificing themselves so the planets could be made. Our planet also seems to have been formed from such objects, poor in ice, rich in rock. Many other comets were gravitationally ejected from the solar system altogether as, sooner or later, they made close passes with the jovian planets, and especially Jupiter. But the calculations show very clearly that a substantial population of the original comets must have been ejected to the far reaches of the solar system, where the random gravitational shuffling of passing stars would have

It appears reasonable to suppose that during the phase of the formation of the major planets there were besides the planets many small *condensations, which may have had comet-like structures... many of these stray bodies must ultimately have been absorbed by the planets (or proto-planets), but it would be unavoidable that during this period of conglomeration a number of the small condensations suffered large perturbations, bringing them into orbits of considerable eccentricity. The perturbations at subsequent perihelion passages would then start a process of diffusion... in the outward direction... The minute stellar perturbations required to bring the perihelion outside the region of the major planets do not... contribute appreciably to the escape of comets. Their main effect is to... 'catch' them semipermanently in the large cloud surrounding the solar system... Once the stellar perturbations had deviated the orbit so that they no longer pass through the inner part of the solar system, evaporation would practically cease, and the comets could easily retain their volatile components up to the present time.*

J.H. Oort,
'Empirical Data on the Origin of Comets' Chapter 20, in G.P. Kuiper and B.M. Middlehurst (eds), *The Solar System*, Volume 4.

12.12 *Iapetus, one of the outer moons of Saturn. The Earth aside, as we go further from the Sun, evidence for organic matter in the solar system seems to increase. The bright material on the surface of Iapetus is known to be almost pure water ice; the dark material is thought to be a stain of complex organic matter, the origin of which is still under debate. (Courtesy, NASA.)*

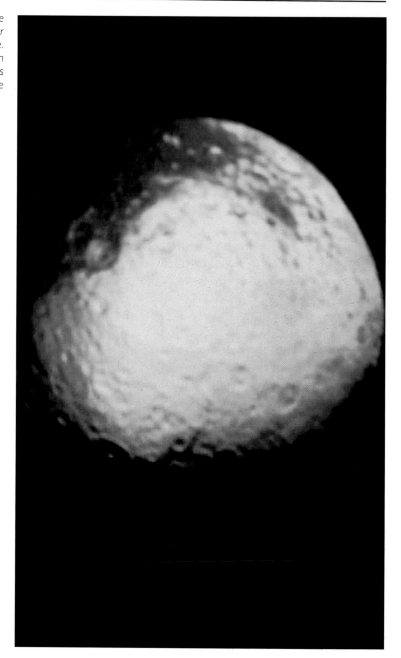

forced them into circular, randomly-inclined orbits. Not all would have been ejected out to the very periphery of the solar system, and the calculations predict a substantial population of comets on near-circular orbits from hundreds to tens of thousands of Astronomical Units out – a population of comets fairly impervious to gravitational disturbances by passing stars. It is therefore possible that a typical denizen of the inner Oort Cloud has never been seen by astronomers on Earth. It is entirely plausible that much bigger comets than those a kilometre across were ejected into the Oort Cloud. But there are far fewer of these, and much more rarely will we see one redirected into our small but well-lit volume of space.

If this currently popular picture is correct, a typical short-period comet is an aggregate of interstellar matter condensed during the origin of the solar system almost five billion years ago – having been ejected by the newly-formed planets, Uranus or Neptune, to the solar system frontiers, its orbit there circularized by gravitational encounters with passing stars. A few billion years later, the cumulative gravitational influence of further stars and interstellar clouds would have driven the comet back into the planetary part of the solar system, where close planetary encounters – this time especially with Jupiter – would have reduced the large elliptical orbit into the more modest dimensions of a short-period comet. The homecoming has been long-delayed, and the solar system has changed considerably in the interim.

Like everything else of which we have evidence, comets are born, live for a time, and then die – or at least disappear. When a short-period comet has wended its long way from Uranus to the Oort Cloud to Jupiter, what happens next? Each time the comet passes through the inner solar system, it runs a gauntlet of risks. Eventually, the odds catch up. For some, every perihelion passage shrinks the comet by a metre, until in the end there is hardly anything left. Other comets collide with something in their path, transmogrify themselves into a different world, or set out for the interstellar void. These several fates have, it turns out, deep consequences for the planets today and, it seems likely, for ourselves. We trace these connections in the following chapters, and a graphic summary of the evolution of the solar system and the births and deaths of the comets is provided on pages 192 and 193.

Because of these various ways for a comet to die, there would, after a while, be no short-period comets at all – if, that is, new comets were not being recruited into the inner solar system by Jupiter's gravity. As on Earth, the places of those recently departed are taken by a new, exuberant, but relatively inexperienced generation.

The comets are stations in the evolution of planets from interstellar matter. They have seen much. As remnants of the forming solar system, sequestered for billions of years in mint condition, they can also tell us much. This is the principal motivation for the dawning age of exploration of the comets by spacecraft. Both the comets and the planets are formed of interstellar materials. The difference is that the planets have been enormously reworked, both physically and chemically, since the beginning of the solar system, while the comets remain comparatively untouched by the ravages of time. When we study the comets, we study our own beginnings.

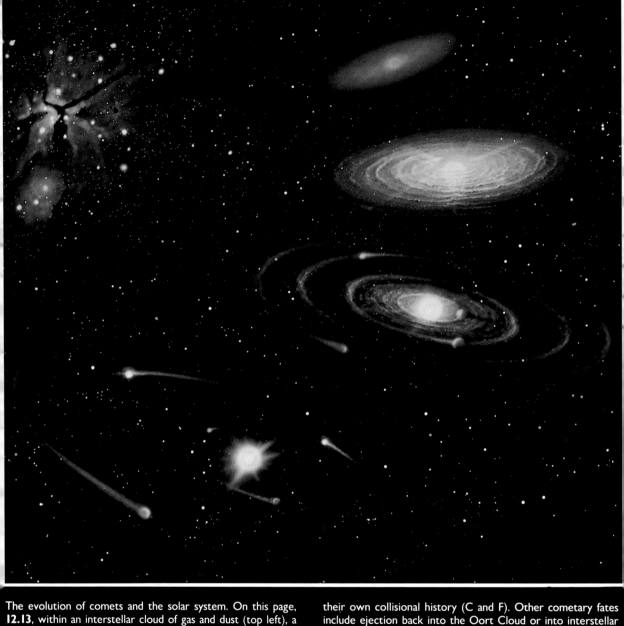

The evolution of comets and the solar system. On this page, **12.13**, within an interstellar cloud of gas and dust (top left), a solar nebula (top right) forms and evolves. The collision of kilometre-sized condensations and subsequent collisions and ejections eventually produces a solar system with a small number of planets (bottom) and a large number of comets (not shown). (Sequence of paintings by Kazuaki Iwasaki.)

The evolution of the residual small debris – objects 100 kilometres in size and smaller – is shown schematically on the facing page. A cometary nucleus (A) transported to the inner solar system becomes a comet with a developed tail (D). Some comets collide with moons and planets producing craters which are also produced by rocky asteroidal objects that have survived

their own collisional history (C and F). Other cometary fates include ejection back into the Oort Cloud or into interstellar space (B); evaporation of the topmost layers of ice, converting the cometary nucleus into an apparently rocky asteroid (E); nearly complete vaporization of ices, so that the comet falls to pieces and becomes a meteor shower (G). Cometary debris falling on the Earth may have made some contribution to the origin of life (H). (A, painting by Michael Carroll; B, diagram by Jon Lomberg/BPS; C, painting by Don Dixon; D, photograph of Comet West, Courtesy, International Halley Watch; E, painting by Don Dixon; F, Apollo 11 photograph of the far side of the moon by Michael Collins, Courtesy, NASA; G, painting by Don Dixon; H, diagram of DNA by Jon Lomberg/BPS.)

A

B

C

D

E

F

G

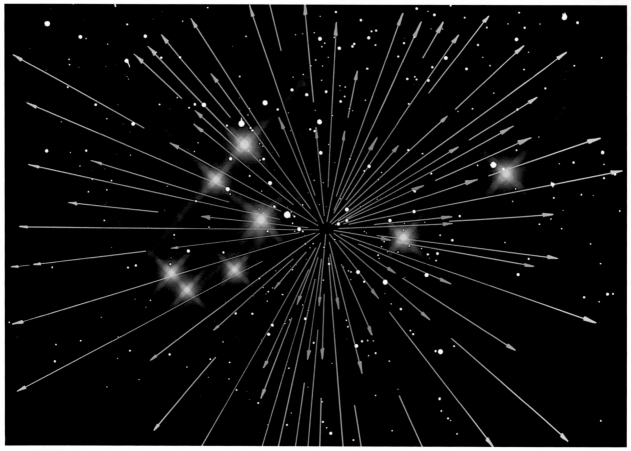

Chapter 13

THE GHOSTS OF COMETS PAST

Oft you shall see the stars, when wind is near,
Shoot headlong from the sky, and through the night
Leave in their wake long whitening seas of flame.

Virgil,
Georgics Book I.

You hear about them before you ever see them. 'Did you see the falling star last night?' the adults ask one another. You wonder how old you'll have to be before they'll let you stay up late and see for yourself. Falling stars. The words conjure up something tragic – a star, high and proud for years, and then for some secret transgression, humbled before our eyes. Cosmic justice.

When you're older, ten say, and you finally get to see your first falling star, you're pleased; it's like fireworks. You might even ooh and ah. You try to remember back, and see whether any star that used to be up there is now missing. But it's hard to do. There are so many faint stars, Even so, you wonder, with meteors falling every night, why there are any stars left at all.

The very phrase, falling star, has a model of the universe built into it. Stars can be loosened from their attachments to the firmament and plummet to Earth. Stars, therefore, must be little things. You see the point of light streaking from the horizon towards the zenith, brightening and then fading. Where did it go? It seems more natural when it falls the other way, from zenith towards horizon. Then you might be tempted to look for a fallen star – by taking a brisk walk over to where the trail of light seemed to end – in the next county, maybe. At age ten, what do you imagine you might find? Something with five points, wrapped in silver paper, glistening in the snow? Perhaps.

The terms 'falling star' and 'shooting star', are, of course, not scientific language. The astronomical word is meteor. A meteor is an object falling through the Earth's atmosphere that produces a streak of light. It looks something like a magnesium flare. If it falls in solitary splendour, it is called a sporadic meteor. If it is a member of a group of meteors all

13.1 *(Opposite) Meteor radiant in Leo. In a meteor shower, most of the meteors (represented here as arrows, indicating the direction of their fall through the Earth's atmosphere) seem to originate in a particular spot in the sky – here within the Constellation Leo, whose stars are indicated in blue. It occurs when the Earth, at that moment carried by its orbit around the Sun in the direction of the Constellation Leo, is ploughing through a cloud of meteoric debris. Other meteor showers have radiance in different parts of the sky and occur on different days of the year. (Diagram by Jon Lomberg/BPS.)*

Through the tranquil and pure evening skies, a sudden fire shoots from time to time, moving the eyes which were steady, and seems to be a star which changes place, save that from the region whence it was kindled nothing is lost, and it lasts a short while.

Dante Alighieri,
Paradise, Canto XV.

And if there are several streams of meteors, which come across that little line in space which constitutes the Earth's orbit, what untold multitudes of them must be within the whole length and breadth of the solar system! Perhaps it may even turn out that the mysterious zodiacal light which attends the Sun, is due to countless hordes of these little bodies flying in all directions through the space that lies within the Earth's orbit.

G. Johnstone Stoney, 'The Story of the November Meteors', Friday Evening Discourse, The Royal Institution, (14 February 1879).

The night was fine; the moon was absent. The meteors were distinguished not only by their enormous multitude, but by their intrinsic magnificence. I shall never forget that night. …For the next two or three hours, we witnessed a spectacle which can never fade from my memory. The shooting stars gradually increased in number until sometime several were seen at once. Sometimes they swept over our heads, sometimes to the right, sometimes to the left, but they all diverged from the East. As the night wore on, the constellation Leo ascended above the horizon, and then the remarkable character of the shower was disclosed. All the tracks of the meteors radiated from Leo. Sometimes a meteor appeared to come almost directly towards us, and then its path was so foreshortened that it had hardly any appreciable length, and looked like an ordinary fixed star swelling into brilliancy and then as rapidly vanishing. Occasionally luminous trains would linger on for many minutes after the meteor had flashed across, but the great majority of the trains to this shower were evanescent. It would be impossible to say how many thousands of meteors were seen, each of which was bright enough to have elicited a note of admiration on any ordinary night.

Account of the Leonid meteor shower of 13 and 14 November 1866, from Robert Ball, *The Story of the Heavens* (London, Cassell and Company, Revised Edition, 1900).

falling on the same night from the same part of the sky, it is a part of a meteor shower. The brightest meteors are called fireballs, and a very large fireball can be as bright as the Moon or even the Sun. Its brilliant head is shaped like a teardrop and accompanied by a trail of light that scatters sparks. After a daylight fireball has fallen, a trail of dark smoke is sometimes seen. By definition, a meteor does not strike the ground. If you follow the streak through the sky, then rush over to the next county and recover a rock newly-fallen from the sky, it is not a meteor, but a meteorite. The diminutive suffix suggests that it is smaller than a meteor, which, in general, is untrue. That big hole in the ground in Arizona is called Meteor Crater – but the object that dug the crater was much too big to be a meteor. Unlike the comets with which they are sometimes confused, meteors, of course, do streak across the sky.

'The meteors whisper greenly overhead', was a beautifully evocative line by the American writer, Loren Eiseley. In fact, meteors whisper only to themselves. They streak too silently through the upper air to be heard down here. Like comets, and children in Victorian dramas, they are seen but never heard. Meteorites – pieces chipped off the asteroids, or the Moon, or Mars, or extinct comets – can be heard; they and some cometary fireballs occasionally produce a deep rumbling roar, the only sound made by another celestial body that unaided Earthbound human ears have ever heard.

Pre-scientific cultures held meteors and meteorites – like comets – to be portents of something, usually impending evil. More rarely, a less anthropocentric explanation is preferred. In West Africa, there are traditions which hold meteors and meteorites to be a kind of solar excrement. Like the notion of a meteor as a star falling to Earth, this teaching of the Atakpame people has more than a grain of truth to it, as we shall see. The Herero call them 'buzzing stones', which doubtless reflects some direct experience with a meteorite fall. In other traditions, the meteors are the souls of the dead returning to Earth to be reborn; or they are a thunder axe; or the heralds of Mbomvei, the supreme being. The Jukun hold that a meteor represents a gift of food carried from one star to another. According to the Kamba, a meteor is a kind of royal ensign, signifying that the beings who live on the stars are this day visiting the Earth. In Islamic Africa, a shooting star was described as a dagger thrown by angels to thwart those genies who aspire to ascend to heaven. But, in general – in Africa and throughout the world – meteors, like comets, were considered portents of pestilence, disaster, witchcraft, and death. However, perhaps because meteors are so much more common than comets – if you're patient, you can see them fall on any clear night – the evils attributed to meteors tend to be more humdrum than those for which comets are held accountable.

From early times, the Chinese (of course) kept meticulous records of meteor showers, with careful attention to colour. The earliest known description, 'Stars fell like a shower', is in the *Ch'un Ch'iu*, the *Spring and Autumn Annals*, describing an event that happened on 23 March 687 BC. In ancient and medieval Europe, careful record-keeping on meteors was virtually unknown, but the occasional bright fireball was considered noteworthy. For example, in the year 1000 – which had been widely advertised to be the date of the end of the world – a contemporary account relates:

> The heavens opened, and a kind of flaming torch fell upon the Earth, leaving behind a long track of light like a path of a flash of lightning. Its brightness was so great that it frightened not only those who were in the fields, but even those who were in their houses. As this opening in the sky slowly closed, men saw with horror the figure of a dragon, whose feet were blue and whose head seemed to grow larger and larger.

Modern scientific interest in meteors was prodded by the

Daimachus, in his Treatise on Religion *... says that ... for seventy-five days continually, there was seen in the heavens a vast fiery body, as if it had been a flaming cloud, not resting, but carried about with several intricate and broken movements, so that the flaming pieces, which were broken off by this commotion and running about, were carried in all directions, shining as falling stars do. But when it afterwards came down to the ground in this district, and the people of the place recovering from their fear and astonishment came together, there was no fire to be seen, neither any sign of it; there was only a stone lying, big indeed, but which bore no proportion, to speak of, to that fiery compass. It is manifest that Daimachus needs to have indulgent hearers.*

Lysander,
Plutarch (c 46 to c 120 AD).

13.2 *A meteor falls to Earth. (From Amédée Guillemin,* The Heavens, *Paris, 1868.)*

13.3 *An eye-witness rendition of the Great Leonid Meteor Shower of 13 November 1833. (After Fletcher Watson,* Between the Planets, *Harvard University Press, 1956.)*

following account by the German scientist, Alexander von Humboldt, in Brazil on the night of 11 November 1799:

> From half after two in the morning, the most extraordinary luminous meteors were seen in the direction of the East. M. Bompland, who had risen to enjoy the freshness of the air, perceived them first. Thousands of bolides and falling stars succeeded each other during the space of four hours... From the first appearance of the phenomenon, there was not in the firmament a space equal in extent to three diameters of the Moon, which was not filled every instant with bolides and falling stars... All these meteors left luminous traces from five to ten degrees in length... the phosphorescence of these traces, or luminous bands, lasted seven or eight seconds. Many of the falling stars had a very distinct nucleus, as large as the disc of Jupiter, from which darted sparks of vivid light... The light of these meteors was white, and not reddish... The phenomenon ceased by degrees after four o'clock, and the bolides and falling stars became less frequent; but we still distinguished some to North-East by their whitish light, and the rapidity of their movement, a quarter of an hour after sunrise. This circumstance will appear less extraordinary, when I mention that in broad daylight, in 1788, the interior of the houses in the town of Popayan was brightly illuminated by an aerolite of immense magnitude...

Humboldt determined that many observers, including some in Europe, had seen the same marvel that night, and concluded that a meteor shower was something that happened over a large area of the Earth high in the atmosphere. But this straightforward conclusion posed numerous problems:

> Whatever may be the origin of these luminous meteors, it is difficult to conceive an instantaneous inflammation taking place in a region where there is less air than in the vacuum of our air pumps... Does the periodical recurrence of this great [meteor shower] phenomenon depend upon the state of the atmosphere? Or upon something which the atmosphere receives from without, while the Earth advances in the ecliptic? Of all this we are still as ignorant as mankind were in the days of Anaxagoras.

It is a remarkable fact that meteor showers recur on nearly the same date of every year – on 12 August, say, or 14 December – although it might take a day or two to reach maximum intensity and a day or two to fall off. You can go out on a clear night and count the numbers of bright meteors. With an intense shower, you might count dozens every second. You note where the meteors appear to be coming from. They are not randomly distributed over the sky, but are concentrated instead towards a particular place in a particular constellation (see the frontispiece to this chapter). This focus from which the meteors appear to be radiating is called the radiant. As the

13.4 *A meteor shower as depicted in Camille Flammarion's* Astronomie Populaire *(Paris, 1880). The date is listed as 27 November 1872; if 27 is a misprint for 17, this would be the Leonid Shower.*

constellation rises and sets, the radiant moves with it. In fact, the meteor shower is characterized by the constellation from which it seems to emerge. The Leonid shower, on 17 November, arises in the constellation Leo; the Taurids, on 5 November, in the constellation Taurus. Now how, the nineteenth-century astronomers asked themselves, could meteor showers know what day of the year it was, and how did they manage this conjurer's trick of pouring out of a tiny spot of sky, which rises and sets with the stars?

In a meteor shower, the trails all diverge from the same point in space. This is a circumstance familiar to those who must drive in snowstorms at night. As the automobile plunges through the falling snowflakes, they speed by on all sides, apparently diverging from a fixed point just ahead.* Similarly, a meteor shower occurs when the Earth, in its swift annual motion about the Sun, passes through a swarm of fine particles.

*This would also be true for hypothetical interstellar comets attracted by the Sun's gravity; they would most often seem to be arriving from the direction in space towards which the Sun is moving, in the constellation Hercules. The fact that comets do not show a preference for Hercules is one of several powerful arguments against Laplace's proposal that the long-period comets come to us directly from beyond the solar system.

What is happening was described vividly at the turn of the last century by the British astronomer Robert Ball:

> Let us imagine a swarm of small objects roaming through space. Think of a shoal of herrings in the ocean, extending over many square miles, and containing countless myriads of individuals... The shoal of shooting stars is perhaps much more numerous than the herrings... The shooting stars are, however, not very close together; they are, on an average, probably some few miles apart. The actual bulk of the shoal is therefore prodigious; and its dimensions are to be measured by hundreds of thousands of miles.*
>
> The meteors cannot choose their own track, like the shoal of herrings, for they are compelled to follow the route which is prescribed to them by the Sun. Each one pursues its own ellipse in complete independence of its neighbours... We never see them till the Earth catches them. Every 33 years the Earth makes a haul of these meteors just as successfully as the fishermen among the herrings, and in much the same way, for while the fisherman spreads his net in which the fishes meet their doom, so the Earth has an atmosphere wherein the meteors perish. We are told that there is no fear of the herrings becoming exhausted, for those the fishermen catch are as nothing compared to the profusion in which they abound in ocean. We may say the same with regard to the meteors.

To the question of how there come to be shoals of meteors passing by the Earth, an answer was supplied when meteor showers were discovered to be associated with comets – especially defunct comets such as Comet Biela (see Chapter 5). There is now good evidence that fine particles abound in the dust tails and comas of comets. Those that are not removed by radiation pressure and the like continue to move around the Sun as separate microplanets, sharing essentially the same orbit as the parent comet. Because of jetting and the successive shedding of comas, some particles will be moving a little faster and some a little slower than the cometary nucleus. As a result, the particles have slightly different periods of orbit around the Sun. A particle somewhat slower than the rest will trail a little bit after one perihelion passage, more after two, and so on. Eventually, the particles would be spread out over the entire length of the orbit, dispersing a little in lateral dimensions as well. Also, small particles feel radiation forces to which big particles are immune, and the gravity of nearby

---

*Ball reasoned something like this: suppose the peak intensity of a meteor shower lasts about an hour. If we knew how fast the Earth is moving through the swarm, we could then calculate the distance from the front of the swarm to the back. The Earth is 150 million kilometres from the Sun (93 million miles) and so its entire circular orbit is $2\pi \times 1.5 \times 10^8$ kilometres around. It takes 365 days for the Earth to make this circuit. So, the Earth travels at about $2\pi \times 1.5 \times 10^8$ kilometres/365 days = 2.5 million kilometres per day. This is the speed of the Earth around the Sun. Thus, if the meteor shower lasts a day, the swarm must be 2.5 million kilometres across. Since, generally, the peak intensity lasts only a few hours, the inner core of a typical meteor swarm is a few hundred thousand kilometres across.

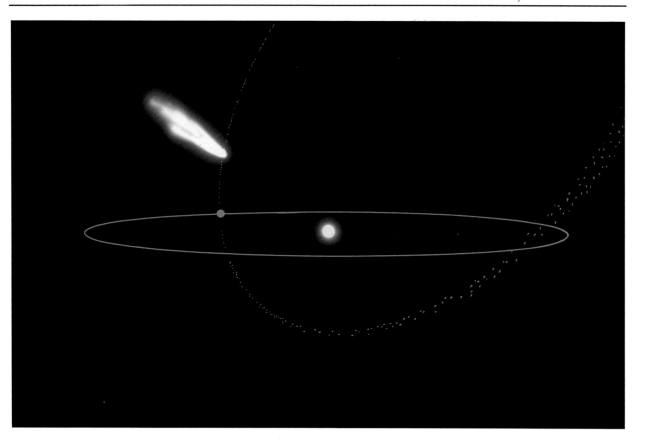

planets, too, serves to spread the stream of particles. A swarm of small meteors can travel round the Sun, shoulder to shoulder in the same orbit, with virtually no collisions. As a comet slowly dissipates, it fills its entire orbit with debris.

In most cases, the shoals of meteors in a cometary orbit are entirely invisible to astronomers from Earth. But, occasionally, by chance, the orbit of the comet intersects the orbit of the Earth. Since the Earth is in a specific segment of its orbit on every day of the calendar, the resulting meteor shower must occur on a specific day of the annual calendar. The main meteor streams today are the Perseids, the Leonids, the Orionids, and the Geminids; also, the Taurids, associated with the periodic Comet Encke. Every comet litters its orbit with debris when it dies. Some orbits cross the Earth's. Meteor showers are the ghosts of comets past.

Just as Tycho Brahe was able to determine from observations of parallax that comets passed far beyond the Moon (see Chapter 2), so photographic observations of the same meteor, made by separated cameras on the Earth, can determine how high up the meteorite trail is. The typical answer is somewhere in the vicinity of 100 kilometres (sixty miles). At this altitude, the atmospheric pressure is 0.00003 percent of that at the surface of the Earth; Humboldt was

13.5 *A still active comet, through successive perihelion passages, has begun to break up. Fine cometary debris now litters its entire orbit, spreading out from the width of the original cometary orbit as well. The comet itself passes well behind the Earth's orbit in this diagram, but the cometary debris intercepts the Earth's orbit. Thus, on a given date of each year, the Earth runs into the cometary debris, producing a meteor shower. (Diagram by Jon Lomberg/BPS.)*

13.6 *The Earth does not pass through most meteor streams left by decaying comets. In the few cases where the debris in the cometary orbit intersects the Earth's orbit, meteor showers occur. Here, small segments of the meteor-strewn orbits of seven comets are shown where they intercept the orbit of the Earth. Each intersection, for the debris from a different comet, corresponds to a particular date in the year. The Earth itself is shown in blue. (Diagram by Jon Lomberg/BPS.)*

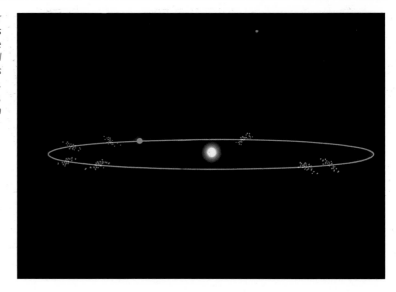

entirely right to wonder how so brilliant a trail could be made by passing through air so thin.

Imagine you are holding a piece of comet stationary above the Earth, and then let it go. It speeds up, of course, as it falls. By the time it reaches the Earth's upper atmosphere, it will be moving at escape velocity – about eleven kilometres a second (seven miles a second). Ordinarily, a comet will have some velocity relative to the Earth before it is attracted by the Earth's gravity, and will therefore hit it at a higher speed. A comet on a highly eccentric orbit, moving in a retrograde direction so it collides with the dawn hemisphere, can be travelling as fast as seventy-two kilometres per second. By contrast, the typical muzzle velocity of a rifle bullet is about one kilometre per second. When a meteor enters the Earth's atmosphere, it is heated to incandescence by friction with the thin air at an altitude of around a hundred kilometres. Spectroscopy of meteors shows spectral lines of iron, magnesium, silicon, and a range of other elements that make up ordinary rocks. Organics, and even some ices, may be in them before they enter the Earth's atmosphere, but at least the silicate component is still there as the meteors burn and die.

Big particles the size of your fist, or larger, streak through the Earth's atmosphere, heat up by friction with the air, and char, melt, or burn off a thin crust. This process protects the interior of the meteorite, just as the ablation shield on a spacecraft protects astronauts during re-entry. The remainder of the object survives the passage through the Earth's atmosphere and reaches the ground, where, when they are recovered, they are called meteorites.

Very small particles are able to radiate their heat away quickly – because they have such a large area for their mass –

13.7 *(Opposite) A swarm of cometary debris falling into the Earth's atmosphere. Some of the particles float down, some make it all the way to the surface as large pieces of matter, and particles of intermediate size burn up on entry as meteors. (Painting by Don Dixon.)*

Fig. 49.—Bolides.

13.8 *A representation of meteors melting and burning up as they enter the high atmosphere of the Earth. (From H.W. Warren, Recreations in Astronomy, 1879.)*

and so they never melt. They simply slow down at a hundred kilometres of altitude, where they contribute to the rare noctilucent clouds that reflect sunlight back to the Earth at night. Gently, they fall for years through a barrage of bombarding air molecules which tend to keep them suspended. Eventually, they enter the circulation of the lower atmosphere and are carried down to the Earth's surface. They are called micrometeorites.

Particles of intermediate size are too small to survive the charring of even a thin crust, and too large to float down gently. They burn up entirely during entry. These are the meteors.

With radar techniques and a network of fast cameras, it is possible to measure how the meteor decelerates when it enters the Earth's atmosphere, and how it flares. Information about mass and density is derived. Typical visual meteors are millimetre sized – no bigger than a small pea. A fireball as bright as the brightest star typically weighs less than a hundred grams; but a fireball temporarily as bright as the Sun would weigh a hundred thousand tons. A porous object presents more area for air resistance to work on than does a denser object with the same mass, and so decelerates differently. In this way, a wide range of densities is determined for the meteors of different showers. For example, the Geminids are meteors with the density of ordinary terrestrial materials, around one gram for every cubic centimetre. Most meteors, however, have much lower densities. Comet Giacobini-Zinner is the presumed source of the Draconid meteor shower – once formidable, now not very impressive – on the evening of every 10 October. The density of the Draconids is very low, possibly as small as 0.01 grams per cubic centimetre. To maintain such fragile structures, these meteoroids could not have been violently ejected from their parent body. So there seem to be two populations of objects that enter the Earth's atmosphere and produce fireballs – one very much like the meteorites that are recovered; the other very fragile porous structures that do not correspond to many macroscopic objects made on Earth.

With modern photographic and radar techniques, it is possible to calculate the speed and direction from which a meteor comes, and then to extract from this data the meteor's orbit. Sporadic meteors – those unconnected with meteor showers – tend to lie in the ecliptic plane and go around the Sun in the same direction that the planets and the short-period comets do. The shower meteors, on the other hand, have much larger eccentricities and orbital inclinations, although some certainly lie in the ecliptic plane. Here again, a kind of natural selection has occurred. Short-period comets with small inclinations produce meteor streams that tend to be disrupted by Jupiter's gravity, and fragments in a range of orbits are produced – some of which become sporadic meteors. But comets with large orbital inclinations tend to avoid Jupiter, so that the meteor streams they produce tend to remain intact for much longer. Of more than 40,000 meteor trails studied, not a single one has an orbit that originated beyond the solar system.

Only three meteorites have ever been recovered that are fireball remnants: Lost City, Pribran, and Innisfree, each named after the locale near which it was recovered. They are ordinary silicate kinds of meteorites. Those bright meteors that arrive from beyond Jupiter have the stirring name of 'transjovian fireballs'. As determined from their entry characteristics, they are as fragile as the most delicate meteorite known. If a sizeable piece of such material were gently placed on the table before you it would collapse under its own weight. It is possible that the spaces in these silicate dust balls were originally, on the parent comet, filled with ices and organics.

On 5 May 1960, the Soviet Premier, Nikita Khrushchev, made a brief announcement that an American airplane had, four days earlier, violated Soviet airspace and had been shot down. A little later on the same day, the newly-formed US National

13.9 *A lithograph by Honoré Daumier, entitled 'Comet of 1857'. However, the woman's urgency suggests that the comet is 'streaking' across the sky, in which case it would be not a comet but a meteor. In studying meteors through a telescope, since we do not know beforehand the precise position of each meteor, it is much more sensible to use a very wide-angle telescope than the spyglass variety depicted here. (Courtesy D.K. Yeomans.)*

13.10 *The Lockheed U-2 aircraft, designed for surveillance overflights for the Central Intelligence Agency, and now re-dedicated to scientific research by the National Aeronautics and Space Administration. This model flies into the stratosphere to capture cometary debris. (Courtesy, Ames Research Center, NASA.)*

Aeronautics and Space Administration made a related announcement that revealed, for the first time to many, that there was a new kind of aircraft, called the U-2. It could fly very high. A research plane of this sort, in the course of studying 'meteorological conditions at high altitudes', NASA said, had gone missing over the 'mountainous and rugged' area of Lake Van, Turkey. Perhaps it had strayed over the border into the USSR. The airplane was described variously as a 'flying test bed', and a 'flying weather laboratory'. It had been used by NASA, among other purposes, to determine 'the concentration of certain elements in the atmosphere'. State Department spokesman Lincoln White said: '...there was absolutely no deliberate attempt to violate Soviet airspace, and never has been'. Three days later, Mr Khrushchev announced that the U-2 airplane had been shot down over the Soviet Union – near Sverdlovsk, it turned out, more than 2,000 kilometres from Lake Van. Its pilot, named Francis Gary Powers, and some of his photographic equipment had been recovered in one piece. Powers admitted to working for the US Central Intelligence Agency on a daring espionage overflight of the Soviet Union. The plane, it turned out, had no equipment for analyzing the atmosphere, but Khrushchev displayed some of the equipment that Powers acknowledged bringing with him – including a pistol with silencer, a poison capsule to swallow if captured, 7,500 roubles in Soviet currency, French, West German, and Italian money, three watches, and 'seven gold rings for ladies'. Khrushchev asked:

Why was all this necessary in the upper layers of the atmosphere? Or maybe the pilot was to have flown still higher to Mars, and was going to lead the Martian ladies astray. You see how thoroughly American pilots are equipped before setting off on a flight to take samples of air in the upper layers of the atmosphere.

The American response, making no reference to the NASA cover story of only a few days earlier, talked about the importance of obtaining data about Soviet military capability because of the closed nature of Soviet society. The incident is historically important for a number of reasons, including its scuttling of the scheduled summit conference between Khrushchev and the American President, Dwight Eisenhower. It also threatened the integrity of the spanking new National Aeronautics and Space Administration, which at Eisenhower's explicit request was to have been dedicated to peaceful research in science and technology.

The U-2 was designed for photographic reconnaissance from an altitude too high to shoot down conveniently. But as Soviet surface-to-air missiles improved, U-2s were mainly supplanted for intelligence work by reconnaissance satellites. Once they had become obsolete for espionage, U-2s were put to use for science in earnest. It is, therefore, a mildly ironic footnote to history that, many years later, the U-2 was the key for fundamental – even trail-blazing – findings in the upper atmosphere. For the first time in human history, it managed to capture pieces of a comet and bring them home to be examined. The driving force behind this programme has been Donald Brownlee of the University of Washington in Seattle.

A U-2 takes off from NASA's Ames Research Center on the Moffett Field Naval Air Station near Mountain View, California. The aircraft has immense wings for its size; it looks a little ungainly, as if it were a cross between a glider and a jet aircraft. Attached to a pylon mounted on the wing are collecting surfaces of sticky silicone grease. The surfaces are not opened to the airstream until the U-2 reaches an altitude of some twenty kilometres. The aircraft is flown more or less at random, because the pilot cannot know where in the stratosphere concentrations of fine cometary or meteoritic dust may be found. For every hour that it flies, ramming the sticky plate into the air in front of it, the plane collects about one big particle (more than ten microns across, still invisible to the naked eye), and many smaller particles. The plate is then automatically covered, the plane swoops down out of the almost black sky, and the day's catch of stratospheric dust is examined through the microscope back on Earth.

It is extremely rare that a small particle, lifted a little off the Earth's surface, one day finds itself in the stratosphere. Most particles tend to be rained out or carried out before they reach high altitudes. There is a natural barrier to movement between

13.11 *A sticky collecting surface attached to the wing of a U-2 aircraft to gather up cometary debris from stratospheric altitudes. (Courtesy, Ames Research Center, NASA.)*

13.12 *Probable cometary debris collected at stratospheric altitudes and examined under a scanning electron microscope. At the top, we see many fine particles stuck together in a morphology resembling a cluster of grapes. The magnification is of 5,400 times and the small horizontal bar is one micron (one ten-thousandth of a centimetre) across. At the bottom is a close-up of an individual submicron particle, which may just conceivably be a relatively unprocessed interstellar grain. (Courtesy, Don Brownlee and Maya Wheelock, University of Washington, and NASA.)*

the lower and the upper atmosphere. (This restriction is removed in a nuclear war, but that is another story.) So as long as we are somewhat restrained about polluting the atmosphere with fine particles, the stratosphere should act as a natural catchment region for extraterrestrial particles falling to the Earth from space.

You take the sticky plate off the wing of the U-2, and put it under the microscope. You count the number of particles of various sizes, you photograph them, you try to do some chemical analysis – although this is difficult because there are few big particles. For the same reason, you try to avoid destructive testing. There are many other scientists in line to examine this material after you. Between examinations, the particles are stored at the old Lunar Curatorial Facility at Johnson Spaceflight Center in Houston, where the rocks returned by Apollo astronauts from the Moon are also kept.

One kind of dust you find is very simple – roughly spherical particles of pure aluminium oxide, collected at twenty kilometres of altitude. How did they get there? These particles turn out to have been generated by solid fuel rockets – largely of American, Soviet, and French manufacture – as they accelerate through the stratosphere on their way to more distant places. The population of aluminium oxide particles in the stratosphere is growing, but it still represents only a distraction. The sticky plates have captured much stranger particles.

When the most abundant kind of stratospheric dust particle is examined under extremely high magnification, it shows itself to be an irregular aggregate of still tinier particles, individually about a tenth of a micron across: a hundred thousand of them side by side would be as long as the nail of your little finger. There is no known industrial or biological process that makes particles like this. Even if there were, there is no way for such particles to be transported to the stratosphere in such abundance. A range of physical and chemical tests all point in the same direction: these particles have originated on another world. Since the principal source of fine particles in our region of space is the comets, it follows that we have before us the stuff of comets.

We are not sure why the particles stick together. Perhaps there was once (or still is) an adhesive material on their surfaces. You can see at a glance that this is a fragile structure. It will collapse if you put much weight on it. If these aggregates were ever at depth inside some small world, they would have been crushed – unless the spaces had been filled by some other, more volatile material, which has now gone. This is another reason why the stratospheric dust particles plausibly arise from the tails of comets. The ice has now all evaporated, and even some of the organics have probably been lost.

Look at these tiny grains again. They were once on the surface of a cometary nucleus. If you were able to get a close-up view of an undisturbed piece from the interior of a cometary nucleus, you would probably see such clusters of small, dark, organic-rich particles, but with all the voids and surrounds filled with ice. As time passes in the inner solar system, the ice evaporates, the particles fly off into space, and some of them enter the Earth's atmosphere.

Ninety percent of the fluffy particles of probable cometary origin collected in this way are aggregates, as we see in Figure 13.12, of tiny, sub-microscopic dark grains, very likely an intimate mix of silicates and related minerals with complex organics. These are, very roughly, the same size as the grains that fill the spaces between the stars, and it seems just barely possible that we have before us a sample of the original material from which the solar system formed. There is not yet enough of it available to be certain. Much larger scale collection programmes by the U-2 and its successors are needed – both to accumulate enough material to do a variety of physical and chemical tests, and to search for rarer particles which may have something surprising to tell us. Ultimately, we will want to compare these particles with particles actually collected from the nucleus of a comet. But, for now, it seems at least possible that we are looking for the first time at the building blocks of worlds.

Schematic diagram of the evolution of a cometary surface, in extreme close-up. Far from the Sun, the surface is an intimate mixture of ice (shown here in blue) and rocky and organic material (shown here in brown). As the comet approaches the Sun, the temperatures increase until, **13.13** (top left) the ices begin to vaporize. Soon, **13.14** (top right), most of the surface ices have been lost to space until, **13.15** (lower left) only rocky and organic materials are left. The pieces of this composition break off from the comet; when one enters the Earth's atmosphere, **13.16** (opposite), it becomes a meteor or, if quite large, a fireball. (Paintings by Kim Poor.)

ANNE NORCIA 1985

Chapter 14

SCATTERED FIRES AND SHATTERED WORLDS

Often, when the Sun has set, scattered fires are seen not far from it.

Seneca,
Natural Questions
Book 7, 'Comets'.

The angels all were singing out of tune,
*   And hoarse with having little else to do,*
Excepting to wind up the sun and moon,
*   Or curb a runaway young star or two,*
Or wild colt of a comet, which too soon
*   Broke out of bounds o'er the ethereal blue,*
Splitting some planet with its beautiful tail,
As boats are sometimes by a wanton whale.

Lord Byron,
The Vision of Judgment (1822).

George-Louis Leclerc, Comte de Buffon, a naturalist who died on the eve of the French Revolution, seems to have been the first person to attempt to reconstruct, from the record in the rocks, the history of the Earth as a succession of geological ages. He also proposed that in ages past a massive body that he called a comet struck the Sun, launching out into space great blobs of fiery matter which cooled, condensed, and formed the planets and their moons. Laplace soon showed that this reconstruction of events could not account for the orbital regularities in the solar system; but it is the first mention in science of cometary matter hitting the Sun, and the first modern attempt to describe how comets die. In this chapter, we describe four of the many fates of comets: to hit the sun directly; to disintegrate, with the remains spiralling into the sun; to impact the moons and planets; and to be transformed into a different kind of world.

In the nineteenth century – when no one knew that there was such a thing as the nucleus of an atom, much less a science of nuclear physics – it was widely held that meteors make the Sun shine. Meteors must be falling into the Sun, it was argued, surrendering their energy of motion, and thereby heating further the fiery surface – which, in consequence, obligingly radiated light and heat to the needy inhabitants of the Earth.

14.1 *(Opposite) A comet impacts the luminous surface of the Sun. The dark features are sunspots with locally-enhanced magnetic fields. There are no solids or liquids anywhere in the Sun; everything is gaseous. (Painting by Anne Norcia.)*

But meteors are the debris and sometimes the shrouds of dead and dying comets; so if this view were correct, comets would be responsible for life on Earth. In fact, it is hydrogen fusion, not meteor impact, that makes the Sun shine, although comets and their debris regularly strike the Sun and the planets – and may in several senses be responsible for life on Earth.

Tewfik's Comet (1882) came fairly near to hitting the Sun; it is called – one of the more romantic names in cometary science – a sungrazer. It was discovered only when near the Sun during a total solar eclipse. Before perihelion passage, it showed a single nucleus. Afterwards, as it headed out past the planets, it revealed itself to have split into four separate nuclei, which gradually drew away from one another. The Sun tugs on the near side of the comet a little more strongly than on the far side, and for a very fragile structure and a close passage this might be enough to break it into two or more pieces. The computed periods for the return of these components are all between 500 and 900 years in the future, but separated by intervals of about a century. The four fragments – each of considerable size – have been predicted (very roughly) to return in the years 2546, 2651, 2757, and 2841. The generations of observers in those remote times will see four comets, spaced about a century apart, pouring out of the same small region of

14.2 *A comet is discovered near the Sun during a total solar eclipse. (Painting by Anne Norcia.)*

the sky and heading for the Sun. They will, perhaps, briefly recall what they have been taught about life towards the end of the second millennium.

The Comet of 1882 is itself one of the so-called Kreutz family of sungrazing comets that episodically arrive from the same place in the sky. The great sungrazing comets of 1668, 1843, 1880, 1887 (the last, like the ghost seen by Ichabod Crane, apparently headless) are all part of this family, and it is natural to wonder whether they might be fragments of a still larger ancestral comet that, in another age, came too close to the Sun and was torn apart by the solar gravity.

The closer a comet passes to the Sun, the stronger is this disruptive tidal force. Comet Ikeya-Seki 1965 VIII was also a Kreutz family daylight sungrazer, but did not fragment. The Great Comet of December 1680, studied by Newton and Halley, was also a sungrazer, passing within 200,000 kilometres of the Sun's surface – closer than the Moon is to the Earth. It, too, did not split during perihelion passage. On the other hand, Comet West 1976 VI, despite coming no closer than thirty million kilometres from the sun, split into four fragments which slowly receded from each other at more than their mutual escape velocity. Thus, the gravitational tides of the sun cannot be the sole cause of the splitting. Except for the sungrazers, we still do not know why comets split.

14.3 *Comet West: a set of four different exposures taken on each of three different nights in 1976. On the long time exposures, the coma is brought out but details in the inner coma are obscured. As the comet split, the components moved apart, each varying in brightness from night to night, and each exhibiting a separate tail. The photographs were taken at the 154-centimetre instrument on Mount Lemon. (Courtesy, Steven Larsen, University of Arizona.)*

EXCLAMATION MARK

Scientific papers tend to be low key, and the appearance of an exclamation mark – except in mathematical papers, where it stands for something very different – is almost unknown. Here is one of the exceptions. Brian Marsden, of the Center for Astrophysics in Cambridge, Massachusetts, is the Director of the International Astronomical Union's office that alerts astronomers all over the world to the discovery of a new comet, and to much else. Marsden himself has made a number of important contributions to understanding the comets. In a scientific paper,* published in the 1960s, he is discussing the sungrazing comets, and finds that there are two members, 1882 II and 1965 VIII, that look as if they had split apart near the aphelion. But aphelion is disconcertingly far away:

> Although most of the comets observed to split have done so for no obvious reason, one really does require an explanation when the velocity of separation is some 20% of the velocity of the comet itself! A collision with some asteroidal object at 200 AU from the sun, and 100 AU above the ecliptic plane, even though it would only have to happen once, is scarcely worthy of serious consideration.

The problem is left unsolved.

*B.G. Marsden, 'The Sungrazing Comet Group' *Astronomical Journal*, Volume 72, page 1170 (1967).

14.4 *Frames from a movie shot in space: Comet Howard-Koomin-Michels impacts the Sun on 30 August 1979. (Courtesy, Naval Research Laboratory, Washington, DC.)*

Splitting and jetting may be connected (see Figure 9), Chapter 7). At the moment Comet West split, the individual fragments brightened noticeably, and propelled large quantities of dust into space in the first of some dozen bursts. Eighty percent of comets that split do so when they are far from the Sun in the vicinity of the asteroid belt. Comet Wirtanen fragmented in 1957 as it passed the orbit of Jupiter. Comet Biela is a similar case. Such splitting is probably due to the vaporization of exotic ices. These sungrazers have all passed within about half a million kilometres of the Sun's surface, penetrating through the thin, hot gas of the solar corona, where the temperature hovers around a million degrees. But the comet spends so little time in the solar corona, and the tremendously hot gas there is so thin, that it must be the Sun's gravity, more than its heat, that is responsible for the fragmentation of the sungrazers.

If some comets come so close to hitting the Sun, should not others fall in and impact it head on? Occasionally, comets approach the Sun, are lost in its glare, and are never seen again; but perhaps they have only been tidally disrupted into many small pieces by the Sun's gravity. However, through an entirely chance discovery, we can now be sure that comets do sometimes hit the Sun – as Buffon, on a larger scale, proposed.

The US Naval Research Laboratory flies a video camera and telescope piggyback on an Air Force satellite, to monitor the changing activity of the solar corona – perhaps it is an alarm system for solar flares and other storms of charged particles from the Sun, which may damage instruments and astronauts in space (see Chapter 9). A small opaque disc is positioned in front of the telescope so it blocks out the Sun; the fainter corona can now be photographed. At the end of August 1979, this instrument made an extraordinary discovery, which, however, remained unreported for two and a half years, 'owing to delays in release of the data tapes to experimenters'. It obtained, by accident, a time-lapse motion picture of a comet on an impact trajectory with the Sun. The comet, called Howard-Koomin-Michels 1979 XI, turned out also to be a probable member of the Kreutz family of sungrazing comets we mentioned earlier. The comet was headed around the Sun, but its perihelion distance was too near. It never emerged on the other side of the Sun, and must have vaporized completely. Imagine the comet streaking through the solar corona, madly releasing gas, tumbling erratically, fragmenting, grains vaporizing in a staccato of small flashes until, its substance spent, all that remains is a cloud of atoms – mostly H and O – stirred in the white-hot hurricane that constitutes the outer, cooler layers of the Sun. The debris, transported by the solar wind, brightened the corona for hours.

We do not know how many other sungrazers have suffered a similar fate. Comet Howard-Koomin-Michels had not previously been known to exist. It seems unlikely that more than one comet hits the Sun each year; but even this would correspond to about ten times more mass falling into the Sun every year from intact comets than from interplanetary dust. Over the age of the solar system, this might be barely enough to make the dying Sun, billions of years from now, shine on for one more decade. There may be beings, resident on our planet then, who will be grateful for a decade's grace.

Find a place, if you can, far from the city, far from atmospheric pollution, far from artificial light. Pick a clear and moonless night, preferably in springtime in the early evening. Look near where the Sun has set. Direct sunlight is now blocked by the western horizon, and you can make out a more feeble light. If you live in northern mid-latitudes, you'll see a faint luminous band rising upwards and to the left from the place where the Sun has set. It is not the Milky Way, and is only a little harder to find. Seneca called it 'scattered fires', and that is what it looks like – distant flames in, or reflections off, the sky.

If you observe in very clear air, especially at tropical latitudes, you can detect this band of light sweeping completely around the sky, through the zodiacal constellations among which the planets move; and for this reason, it is called the zodiacal light. Just opposite the Sun is a patch of light, brighter than its surroundings, called the 'Gegenschein' (German for counterglow). From the geometry, it is clear the Earth – with the other planets – is immersed in a flat ring of material enveloping the Sun that reflects its light back to Earth.

This further example of disc-like structures in astronomy did not escape the notice of Immanuel Kant. While Kant had

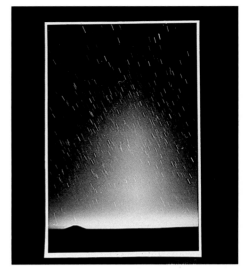

14.5 *A photograph of the zodiacal light extending almost 50° above the horizon. This is a time exposure (note that the stars are shown not as points but as short trails) taken about one hour after sunset by William K. Hartmann at the Planetary Sciences Institute, Tucson, Arizona.*

14.6 *A longer time exposure of the zodiacal light (notice the longer star trails – the north celestial pole, about which the sky seems to revolve, is just off the upper right-hand corner of the picture). To the naked eye, of course, the zodiacal light is much fainter. (Courtesy, William K. Hartmann, Planetary Sciences Institute, Tucson, Arizona.)*

more than one view on the matter, his description of the nature and origin of the zodiacal cloud is memorable:

> The Sun is surrounded with a subtle vapour which surrounds it in the plane of its equator, and is spread out to only a small breadth on both sides but extending to a great distance. We cannot be certain whether... it comes into contact with the surface of the Sun... or, like Saturn's ring, is everywhere removed to a distance from it. Whichever of these views is correct, there remains resemblance enough to bring this phenomenon into comparison with Saturn's ring... This necklace of the Sun... may have been formed out of the universal primitive matter whose particles, when it was floating around in the highest regions of the solar world, only sank to the Sun after the completely finished formation of the whole system.

Kant seems to be saying that the zodiacal light is reflected off a flattened disc of small particles deriving from the solar nebula, and only lately arrived in the inner solar system. If this reading is correct, Kant is again a century or two ahead of his time.

When you view the zodiacal light, you are observing the remains of comets (and, to a lesser extent, fine debris from the collisions of asteroids between the orbits of Mars and Jupiter). There isn't much of the stuff. The total mass of all the particles in the zodiacal cloud is only some hundred billion tons, the mass of a single comet three kilometres across. But it is made of fine grains of dark dust, mixes of silicates and organics, typically tens of microns across, and very like the particles recovered by aircraft in the stratosphere (see Chapter 13). If you grind up a small comet into ten-micron pieces, and ignore the ice, you produce a trillion trillion particles. Spread them through the inner solar system and they can reflect a fair amount of light.

But we should not imagine interplanetary space to be full of small zodiacal particles. Hold your thumb in front of your nose and move it towards the Sun. If you were able to continue travelling, thumb first, all the way to the Sun, how many zodiacal particles would you encounter? The answer seems to be only a few. A single comet, during a single perihelion passage, can provide a hundredth of a percent of all the particles in the zodiacal cloud.

These particles are now measured routinely by interplanetary spacecraft. For example, between 1972 and 1974, the HEOS-2 Earth satellite of the European Space Agency searched for dust between 5,000 and 244,000 kilometres from the Earth, and found itself flying through a swarm of fine particles, previously not known about. The swarm must recently have formed from a larger parent body (thought to be Comet Kohoutek), because the lifetime of a coherent cloud of grains is very short. But, in the large, the zodiacal cloud is uniform, homogeneous, without lumps or strands or holes – except when a planet is ploughing through it and leaves a kind of tunnel behind. Spacecraft on their way to some distant planet regularly bump into the stuff. But there's so little of it, and the particles are, in general, about a kilometre apart, so far that they do not impede interplanetary flight.

However, the paucity of this material is puzzling. In a hundred thousand years or so, all the present zodiacal particles in the inner solar system should be provided by the dust tails of comets. The solar system is 4.5 billion years old. So why isn't there much more zodiacal dust? You might very well expect 4.5 billion ÷ 100,000 = 45,000 times more zodiacal dust than there is; in which case, the zodiacal light would be brighter than the planets and the stars. Some biblical fundamentalists have even used this puzzle to argue for a solar system no more than 100,000 – and hopefully (in their view) less than 10,000 – years old, in a brave but hapless attempt to reconcile a literal reading of the Book of Genesis with the findings of modern science.

The great mass of fine particles released from comets over the eons are therefore missing. Where have they gone? They have, it turns out, been devoured by the Sun. Small particles, around a few tenths of a micron across, are lost from comet tails and from the inner solar system altogether, accelerated out into the depths of space by the pressure of sunlight. Larger particles also feel the radiation pressure, although it is not enough to drive them significantly outward. But it does serve to lessen the Sun's gravity, making the particle a little lightfooted. There is another, contrary influence, first described by the British physicist J.H. Poynting.* The particle is hot and, he says, in consequence, emits radiation to space. But because of its motion, it

> crowds forward on its own waves emitted in front, and draws away from those emitted behind, so that there is increase of pressure in front and a decrease behind. Thus, there is a force resisting the motion.

Since now the particle isn't moving as fast as is necessary to balance the Sun's gravity, it falls a little inwards, where it is hotter, moves faster, and generates still more resistance to its motion. Slowly, the particle spirals to the Sun. Poynting calculated that a tiny particle of rock with a radius of ten microns would spiral into the Sun in less than a hundred thousand years. Much larger particles are too big to be pushed around by sunlight, and do not spiral in. This fiery fate of small particles orbiting the Sun is now known as the

*From Poynting's Friday Evening Discourse at The Royal Institution, 'Some Astronomical Consequences of the Pressure of Light', 11 May 1906.

And if there are several streams of meteors, which come across that little line in space which constitutes the Earth's orbit, what untold multitudes of them must be within the whole length and breadth of the solar system! Perhaps it may even turn out that the mysterious zodiacal light which attends the Sun, is due to countless hordes of these little bodies flying in all directions through the space that lies within the Earth's orbit.

G. Johnstone Stoney, 'The Story of the November Meteors', Friday Evening Discourse, The Royal Institution, 14 February 1879).

14.7 *The zodiacal light as observed in Japan by a Mr Jones. (From Amédée Guillemin,* The Heavens, *Paris, 1868.)*

14.8 *The zodiacal light as seen in Europe by a M. Heis. (From Amédée Guillemin,* The Heavens, *Paris, 1868.)*

Poynting-Robertson Effect. (The American physicist H.P. Robertson later made the most general formulation of this phenomenon.)

Poynting went on to describe how particles of various sizes will eventually move in orbits:

> so different that they may not appear to belong to the same system. In course of time they should all end in the Sun. Perhaps the Zodiacal light is due to the dust of long dead comets. It appears just possible that Saturn's rings may be cometary matter which the planet has captured, and on which these actions have been at play for so long that the orbits have become circular.

So a typical zodiacal particle spends about a hundred thousand years before it is gobbled up by the Sun, and essentially all the particles in the zodiacal cloud are replenished in such a period. Thus, the great majority of the particles you see in the zodiacal light left their comets of origin long before recorded history; but almost none of them did so before the evolution of human beings on Earth. All told, about ten tons of interplanetary particles spiral into the Sun every second – 300 million tons a year.

In its annual voyage around the Sun, the Earth runs into particles of the zodiacal cloud, mainly in the dawn hemisphere; more rarely, faster moving debris catches up with the Earth in its twilight hemisphere. The total being accumulated all over our planet is about a thousand tons of dust a day. The number of fine particles collected by stratospheric aircraft – U-2s and others – is about what we would expect for particles in the zodiacal cloud captured by the moving Earth. If this cometary dust had fallen to Earth over its entire history at the same rate that it does today, and if nothing destroyed it after landfall, there would be a thin, dark, powdery layer about a metre thick everywhere on Earth. (If a single large comet were pulverized, and all its debris were spread smoothly over the Earth, a layer of about one percent of this thickness would result.)

These tiny grains have had an epic history. Intermingled with specks of ice, they were floating in the gas between the stars for eons; then caught up in a contracting, spinning interstellar whirlpool that eventually formed the solar system; growing to form cometary nuclei that were promptly ejected into cold storage at the outskirts of the solar system; then, plunging towards the Sun, propelled out from the comet as the ices evaporate; orbiting the Sun as individual micro-planets; and finally, inexorably, spiralling in, until – during its passage through the solar corona – each is turned into a puff of gas.

The constituent atoms – silicon, oxygen, iron, aluminium,

carbon, hydrogen, and the rest – are then diffused through the upper atmosphere of the Sun, eventually being gathered into its internal circulation and carried down deep into its interior. Some of these atoms will be transported to the very core of the Sun, to engage in the thermonuclear alchemy that makes our star shine. But they constitute the most minor of contributions. Every now and then, though, an occasional light beam, an odd photon – perhaps illuminating for an instant the way of a gnat – originates in the belt of distant comets orbiting the Sun and, before that, in the darkness between the stars.

You take even a casual glance at most of the worlds in the solar system, and you discover that they're full of holes. There aren't many analogues that come readily to mind for this sort of battered surface – a fact which says something about the Earth. People once compared the cratered surface of the Moon to Gruyère or Emmenthaler cheese, but that doesn't quite evoke its real appearance – craters upon craters upon craters, down to the smallest crater you can see. Some of these worlds look more like a rain-spattered beach. Drop marbles of different sizes at random in plaster of paris and let the surface set. The craters will be nicely circular, with rims, ramparts

14.9 *A comet or asteroid strikes the Moon forming Crater Tycho (named after Tycho Brahe, see Chapter 2). This crater was formed in the lunar highlands later than many of the other craters there; note the craters on top of craters. (Painting by Don Dixon.)*

14.10 *A photograph of the sparsely-cratered lunar lowlands. The largest sharp crater in this image is Flamsteed (named after Britain's first Astronomer Royal, see Chapter 3). Visible beneath Crater Flamsteed is an almost entirely lava-flooded crater from the earliest history of the Moon. (Lunar Orbiter IV photograph in Oceanus Procellarum. Courtesy, NASA.)*

and, occasionally, a central mountain peak. Some craters will overlap with others. The plaster of paris begins to look very much like the surface of a world.

Some of these impact craters, on the Moon for example, are made by asteroids, but most – especially in the outer solar system – are made by comets. Rarely, a world will overtake a comet. More often, the comet will overtake the world, or crash into it head-on. Like the Sun, the moons and planets collect their share of cometary impacts. The Sun, being made of gas, can retain no impact craters. But worlds with uneroded ancient surfaces remind us at a glance of how many comets have, over the eons, died in fatal collisions.

On 30 June 1908, something fell out of the sky in Siberia and knocked down a forest. No impact crater was found. The responsible agent is thought by some to be a good-sized but still fragile piece of Comet Encke. But unless the impacting object is fairly delicate, it will make an impact crater as it strikes the surface. Because the falling body is travelling very fast, the hole it excavates will be larger than itself. If no subsequent geological event obscures the surface, every crater corresponds to an impact, and almost every impact corresponds to a crater. The surface of a moon or planet is, therefore, a kind of master record of the history of the solar system. If you can decipher it, you can uncover the catastrophes of ages past.

THE VIEW FROM THE GOLF COURSE

Even the impact of a smallish fragment of a comet (or asteroid) with the Earth can have significant effects far away. The day after the Tunguska explosion in Siberia, on 30 June 1908, the following letter was written to *The Times* of London. It was published two days later:

Sir, Struck with the unusual brightness of the heavens, the band of golfers staying here strolled towards the links at 11 o'clock last evening in order that they might obtain an uninterrupted view of the phenomenon. Looking northwards across the sea they found that the sky had the appearance of a dying sunset of exquisite beauty. This not only lasted but actually grew both in extent and intensity till 2.30 this morning, when driving clouds from the East obliterated the gorgeous colouring. I myself was aroused from sleep at 1.15, and so strong was the light at this hour that I could read a book by it in my chamber quite comfortably. At 1.45 the whole sky, N. and N.E., was a delicate salmon pink, and the birds began their matutinal song. No doubt others will have noticed this phenomenon, but as Brancaster holds an almost unique position in facing north to the sea, we who are staying here had the best possible view of it.

Yours faithfully,
Holcombe Ingleby.
Dormy House Club, Brancaster, July 1.

Take, for example, the Earth's Moon. The side that permanently faces the Earth and which we know with the naked eye, has two kinds of terrain – dark, smooth lowlands, and bright, rough highlands. Both highlands and lowlands are cratered, but there are many more craters in the highlands than in the lowlands. Because American astronauts and Soviet robots have brought back samples from nine different locales on the Moon, we know something of the composition and – through radioactive dating – the ages of various provinces of the lunar landscape we see from Earth.

The dark lowlands are made of lava that, about four billion years ago, gushed out of the then-hot lunar interior, wiping out any previous cratering record. Thus, for the lowlands, the earliest master record has simply been lost. The sparse number of lowland craters is just what we might expect if the inner solar system had always been as full of comets and asteroids as it is today. But the highlands have far too many craters for that. Here, craters have been superimposed on pre-existing craters – so that, again, a record of the earliest times has been lost. One way or another, these worlds hide the evidence of their very beginnings, although the record of later catastrophes is intact.

From the story told by the dense cratering in the lunar highlands, it is clear that, in the first few hundred million years of lunar history, there were enormously more impacting objects – comets and asteroids – around than there are in interplanetary space today. The same story is told on world after world as our reconnaissance spacecraft fly out among the planets. On Saturn's moon, Rhea, for example, there has been no geological activity for billions of years, and we are presented with a world cratered to saturation. In the beginning, interplanetary space must have been full of boulders and icebergs that crashed into the forming worlds. Indeed, this is *how* the worlds formed, in violent collisions.

Not all the worlds retain the craters of eons past. On some, as in the lunar lowlands, the traces have been eradicated. Something fills the craters in, or rubs them out, or covers them over. On Venus, there are recent lava flows; on Mars, great sand drifts; on Jupiter's moon, Io, a melt rich in sulphur; and on Enceladus, a satellite of Saturn made almost entirely of ice, something has melted the surface – the cratering record on that world was literally writ on water. On Io, the craters may be wiped out in centuries; on Venus, it may take a billion years. But stretching through the entire solar system, written on top of one another, are the chronicles both of ancient impacts, and of more recent geological processes.

Something similar is true on Earth, where running water rather quickly destroys the craters – so that, unless they are

14.11 *Saturation cratering on the surface of Saturn's moon, Rhea. Most of these large holes in the ground have been made, over billions of years, by cometary impact. (Voyager 1 photograph. Courtesy, NASA.)*

14.12 *A view from Voyager of the Saturnian moon Tethys. Because the moons of Saturn are in the outer solar system, the abundant cratering on them is produced almost entirely by comets. (Courtesy, NASA.)*

Images of the innermost large moon of Jupiter, Io, taken by the Voyager spacecraft in 1979 and reproduced in false colour. There are many dark spots and apparent holes in the surface **14.13** (above). But when examined more closely **14.14** (right) and **14.15** (below), every one of them turns out to be volcanic in origin. Eight or nine active volcanoes were observed during the two Voyager flybys. Cometary impact craters are certainly being produced on the surface of Io, but the volcanic flows are eroding them away in short periods of time. (Courtesy, NASA.)

very recent, they are very hard to find. So-called Meteor Crater, Arizona, is only a few tens of thousands of years old. There are very few terrestrial craters that are much older, and these are generally large, or in geologically inactive regions.

Sometimes, an impact will reveal something of the underlying planetary surface. On Mars, for example, there are craters with splash marks around them, indicating the presence of sub-surface water ice that has been momentarily liquefied in the collision; surface debris is carried outward until the water freezes. The impact of a large comet might even bring water or an atmosphere to worlds that are almost airless. When a comet strikes a gas planet like Jupiter, it sweeps through the upper atmosphere, encounters more and more resistance as it plunges deeper and, somewhere below the visible clouds, breaks up, its material circulating over much of the planet. The stuff of comets is then mixed with the air of Jupiter. But it does not seem likely that any known feature in the clouds of Jupiter can be attributed to a recent collision.

Craters have shapes. Some are perfectly bowl-shaped; others are flat, shallow, gently sloped. The shape of the crater does not much depend on the velocity of the impacting object, provided it hits the ground hard enough. The result is the same as it would be if a large explosion were detonated at the point of impact. The shape of the crater depends on how soft the surface is, and on the fragility of the infalling material.

American astronauts on the lunar surface photographed craters too small to see with the unaided eye. Samples returned to Earth have also revealed a plethora of microcraters. Some must be due to asteroidal dust swept up by the Moon; others to fine particles generated on and sprayed over the Moon by large impacts. We also expect some fraction of the lunar microcraters to arise from cometary dust, excavated puff by puff. The bowl-shaped microcraters seem to be due to cratering by fine, rocky particles – from asteroids or cometary silicates. But there is also a population of microcraters that are extremely shallow for their size, more depressions than bowls. The microcraters can only be understood if they were excavated by low density grains impacting the Moon, and these can only be of cometary origin. Microscopic bits of cometary fluff are abrading the face of the Man in the Moon.

And, by collision roulette, larger comets must also hit the Moon from time to time. A lunar soil sample called 61221 shows evidence of the molecules H_2O, CH_4, CO_2, CO, HCN, H_2, and N_2. Volatiles present in this sample – especially HCN – and absent in other lunar samples are suggestive of a recent cometary impact. In a collision, comets can also impress a pattern of magnetism around the impact area, and this may account for some puzzling magnetic anomalies on the Moon.

14.16 *A radar image of the surface of Venus obtained with the largest radar telescope on Earth, at the Arecibo Observatory of the National Astronomy and Ionosphere Observatory, Arecibo, Puerto Rico. The smaller crater at the centre is Lise Meitner, about sixty kilometres across. There is still debate about whether such craters are of impact or of volcanic origin, but the relative sparseness of craters on Venus shows that the surface is continuously being modified – probably by volcanism, as on Io. (Courtesy, Donald Campbell, National Astronomy and Ionosphere Observatory.)*

14.17 *A comet streaks into the Jovian atmosphere penetrating below the clouds. Because of its size, and the orbits of the Jupiter family of comets, Jupiter receives comparatively many such impacts. (Painting by Don Dixon.)*

Sooner or later, comets in the inner solar system run out of gas. Typically, the amount of water ice that disappears in a single perihelion passage is a few metres (about ten feet). Each time the comet passes by the Sun, it shrinks. Of course, the comet is not pure ice, but an intimate mix of ice and dust. Huge quantities of dust are blown off in the great jet fountains that sometimes play within the coma, and dust is also lifted more sedately by evaporating ice. So a comet made mainly of ice will lose ice and dust every perihelion, until the comet is altogether gone, and nothing is left but a powdery contribution to the zodiacal cloud and, perhaps, an occasional sporadic meteor in the skies of Earth.

But now imagine that the comet has more dust than ice.

After the first perihelion passage, there is a layer of dust left on the surface that geologists would call a lag deposit. Some of the dust has been carried away by the evaporating ice, but not all. The next time the comet approaches the Sun, the dust serves as an insulation for the underlying icy dirt. It is now harder for it to heat up. And if it does heat up, it is harder for the underlying ice to escape – the way is barred by the layer of dust. After a number of perihelion passages, so large a lag deposit may be created that no further ice can be lost to space. No longer does the comet form a coma or a tail. It shuts down, closes up shop, becoming just another small, dark lump of matter in the inner solar system.

Something similar will occur if an icy comet with a rocky core periodically enters the inner solar system. Such a comet would have to be very large – at least tens, and perhaps hundreds, of kilometres across – for it to have melted, vertically segregating materials of different properties – the denser rock on the inside, the lighter ices on the outside. Thus, there are at least two possible ways for comets to evolve into small bodies with surfaces of silicates and organics. But there *is* a category of such small, dark objects, some of them in quite eccentric orbits, called the asteroids. The entrancing suggestion that comets might make a metamorphosis from an icy world to a rocky one was first made by Ernst Öpik.

If the metamorphosis really occurs, then there might be a few comets caught in transition. They would sputter as they approached the Sun, the final bare patches of ice vaporizing and the lag deposit enforcing quiescence everywhere else. There are such comets, and, if you observe assiduously, you can sometimes catch them in their quiescent phase. Some comets show no decline in brightness, even after ten or more perihelion passages, but others tend, after a while, to fade steadily into inconspicuousness. Take Schwassmann-Wachmann 1, for instance, although it may be very far from extinct. There is no question that it releases gasses like a comet: the diffuse coma is, at least on occasion, clearly seen, and even the emission lines of CO^+ have been observed. Schwassmann-Wachmann 1 seems to show more coma activity when it is closer to the Sun than when it is further away. But when it is quiescent, it has the brightness and colour of an RD-type asteroid – red and dark from the complex organic molecules it contains. A number of active comets, observed far from the Sun before substantial comas form, also resemble RD asteroids – including Halley's Comet, seen in 1985 when it was near the orbit of Jupiter. Other intermittently active comets – Arend-Rigaux, for example, or Neujmin – resemble S-type asteroids when quiescent, brighter, greyer, made mainly of silicates and metals. It seems clear that when

Microcraters in rock samples returned from the Moon. Both pictures were taken through the microscope. From Apollo 12, the biggest crater 14.18 (top) is thirty microns across. From Apollo 15, the big crater 14.19 (above) with the raised lips is less than two microns across. (Courtesy, D.S. McKay and NASA, Johnson Space Center.)

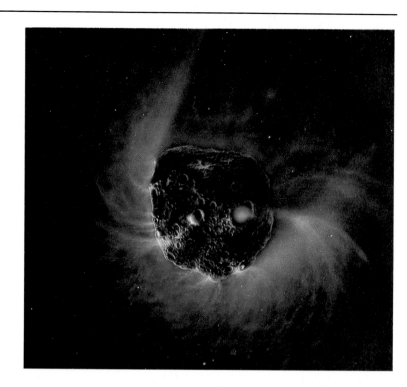

The hypothetical evolution of a short-period comet into an asteroid. An intermittently active comet comes close to the Sun and a small amount of remaining ice near the surface vaporizes **14.20** (above right). The comet then fragments **14.21** (right), exposing previously sequestered ice which also vaporizes. Eventually **14.22** (below), an apparently rocky, ice-free object is left, which is described as an asteroid even if it has an interior still rich in ice. (Paintings by Don Dixon.)

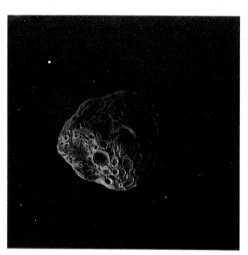

such comets complete their release of gasses, they will be indistinguishable from asteroids.

Asteroids collide with one another, and occasionally a fragment is broken off and finds its way to the Earth's surface, where it often ends its days in a museum, catalogued as a meteorite. There are many different kinds of meteorites. Some have never been heated very much in four and a half billion years, contain complex organic matter, and somewhat resemble the cometary fluff that is collected in the stratosphere. Some of these carbonaceous meteorites could conceivably be of cometary origin, arising from collisions with asteroids. Indeed, the colour and darkness of RD asteroids is like the organic-rich clays in carbonaceous meteorites.

Other meteorites, though, show very little organic content, and are made mainly of stone, or even metal. Some stony meteorites might arise from collisions with asteroids like the one Comet Arend-Rigaux is on its way to becoming. The huge iron meteorites proudly displayed in museums throughout the world tell a story of great heat, melting, and the coalescence of molten iron droplets into a large mass, before the whole thing cooled. In their history, the iron meteorites have been so strongly heated and reworked that they cannot possibly be unaltered, primordial cometary stuff. The same is true, although to a lesser extent, of the stony meteorites. But the carbonaceous meteorites are much closer to the primitive matter from which the solar system formed.

Many of the asteroids originated roughly where they are right now – mainly between the orbits of Mars and Jupiter. They may be the remnants of the rocky (and metallic) bodies that, along with some cometary matter, accreted to form the Earth and the other terrestrial planets. They must be the source of many meteorites. But interspersed among them, travelling incognito and cunningly disguised, are comets. They've put on an overcoat of dust, or stripped down to some hidden core. A few asteroids even have cometary sorts of orbits – highly elliptical, carrying them near Mars, or the Earth, or even Venus. One asteroid of that sort is called Hidalgo, and is of the RD class. Because of perturbations by Jupiter, Comet Encke is slowly walking its orbit inwards, and in a few centuries its orbit should cross that of the Earth. There is abundant evidence that Encke is falling to pieces – including the meteor shower which it trails, as well as the fragment which has broken off and is thought to have been responsible for the Tunguska explosion.

One of them, 1983 TB, was discovered by the Infrared Astronomy Satellite (see Chapter 12) in 1983, and shares its orbit with the meteor stream responsible for the Geminid meteor shower every 14 December (see Chapter 13). It is

Ice and iron cannot be welded.

Robert Louis Stevenson,
The Weir of Hermiston,
(1896).

extremely unlikely that 1983 TB would find itself exactly in the Geminid meteor stream by accident, and it is equally improbable that it is an asteroid that was fragmented by collision just before we started observing it. It is probably a piece of an extinct comet. But the Geminid meteors are known to be of stony, not fluffy material. As with comets like Arend-Rigaux, there is a hint here that comets can metamorphose into true rocky asteroids. Thus, when we find impact craters on the Earth with evidence that the causative agent was made of rock, it may nevertheless have been generated by an extinct comet.

Asteroid Oljato is another comet with an identity crisis. Its highly eccentric orbit, stretching from Jupiter to Venus, is comet-like. It reflects light at visible and at radar wavelengths in an entirely anomalous way; no other object in the solar system looks like Oljato. And it seems to be associated with one of the meteor showers. Oljato has also been detected in an unusual way: in orbit about the nearest planet is an American spacecraft called Pioneer Venus, which, among other duties, is busily recording day-by-day the strength of the local magnetic field. There is a steady background field, and occasional anomalies are noted. About a quarter of the magnetic disturbances that have been detected occurred when Oljato passed very close (within 1.3 million kilometres) to Venus and its attendant spacecraft. Asteroids are not expected to have detectable magnetic fields. But if Oljato is a slightly active comet, generating a thin coma, the coma would be quickly ionized by ultraviolet sunlight and would, in turn, produce an intense compression of the magnetic field carried by the local solar wind; as this compression passed by Venus, an intensification of the local magnetic field might be detected. Oljato's strange spectrum suggests it is something rare, but other evidence strongly implies it is a comet at death's door.

Every year there are a few new recruits to the ranks of the short-period comets, dragooned from longer-period orbits by Jupiter's gravity. Aided by sunlight, gravitational tides and rotation, many of these comets fall to pieces and vanish altogether. Others are ejected from the solar system or run into worlds. But a steady population is metamorphosed into asteroids, with orbits ranging from near-circular to highly elliptical. Over the age of the solar system, the number of comets thus disguised must become quite large.

Many asteroids that cross the Earth's orbit are probably extinct comets. Because they tend to be small and dim, they are hard to find. There are fewer than a hundred known objects that cross the orbit of Mars or Earth, and this is after a decade of intense searching. These approaching asteroids are of particular interest to us, because they represent something with serious and calculable consequences: the high-speed collision of an object a kilometre across with the Earth would represent a major catastrophe, of a sort that must have happened often during the history of our planet. While we may describe the Earth as having been hit by an asteroid, what we are describing may really be a comet in disguise.

From the observed population of comets today, and from the cratering record, it is even possible to estimate how many comets or asteroids, and of which kind, have excavated the craters on a given world. Eugene M. Shoemaker is the founder of the US Geological Survey's Branch of Astrogeology – one of the scientists behind the earliest serious work in America to explore the Moon, responsible for the discovery of many near-Earth asteroids, and the world expert on Meteor Crater, Arizona. Shoemaker combines down-here geology with up-there astronomy. With his collaborator, Ruth Wolfe, he has estimated that something like a quarter of the craters seen on the big moons of Jupiter are produced by the long-period comets. Perhaps another quarter are due to active short-period comets – ones that are still producing copious dust trails, outbursts, changes of brightness and the like. Fully half the craters are dug out by non-cometary bodies, including extinct, short-period comets disguised as asteroids. For the Earth, about a third of recent craters are due to the long-period comets, not very different from the story on Jupiter's moons. But the remainder of the terrestrial craters are due to Earth-crossing asteroids, many of which are extinct comets; here, the known short-period comets make virtually no contribution to cratering. Because their perihelia are near to Jupiter, through the circumstances of their capture, the short-period comets much more often run into the moons of Jupiter than they do the Earth.

When a big comet or asteroid strikes a small world, which sometimes happens, the world may be destroyed, smashed to bits. This almost happened to Phobos, a moon of Mars, and to Mimas, a moon of Saturn. If the object that produced the big crater on Mimas had been somewhat larger, the result might have been what is shown in Figure 14.24. Large comets were abundant in the early history of the solar system and, as they were being incorporated into the growing moons and planets, catastrophic collisions then happened with much greater frequency.

On the Moon and Mars and the large satellites of Jupiter, there is evidence of enormous ancient collisions that produced not craters, but basins hundreds or even thousands of kilometres across. In one promising theory, the Moon arose in a massive impact of some failed planet with the Earth, the

14.23 *A view from a Voyager of Callisto, the outermost of the four big moons of Jupiter. Something like half of these craters have been made by comets. (Courtesy, NASA.)*

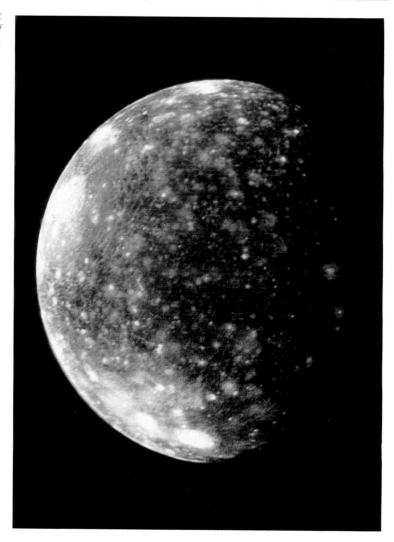

ejected fragments from the collision coalescing to make our natural satellite. The scar has healed long ago, but if that collision had been more energetic, the Earth might have been destroyed altogether, and we would never have come to be.

The spacing between the worlds, their masses, and even their survival, depended on how many primitive bodies were moving on what orbits in the early solar nebula; that is, on the chance concatenation of a multitude of events – some probable, others less so. Start the solar system over again, let only random factors operate, and there will be a slightly different collection of rocky and icy bodies formed around a kilometre across; another sequence of collisions building new worlds; an altered number of planets, with somewhat different masses and orbital positions – but, in general, the rocky terrestrial planets will still be on the inside, and the gas giants on the outside, along with icy and organic moons, and the comets in a great cloud beyond. There may be a vast multitude of planetary systems in the Galaxy that look roughly like our own, but with significant differences from type to type and system to system, depending on the complex collisional roulette that shaped their origins.

Comets rework the surfaces of worlds, carry volatiles to parched planets, dig up buried treasures for the planetary scientists, and leave a record of their history throughout the solar system: they are busy, useful, obliging bodies. But comets also make worlds and destroy them. They remind us of the time some 4.5 billion years ago when the solar system was made of nothing but comets and their rocky equivalents, hundreds of trillions of them or more, swarms of bodies merging and smashing, careening from one district of the solar system to another, a whirlwind of small worlds that eventually led, after the impetuosities of youth had subsided, to the present staid machinery of the solar system. The occasional passing comet is, these days, one of the few reminders of our violent and chaotic origins.

14.24 *The fragments of a shattered Mimas after a cometary impact. Mimas is the innermost of the moons of Saturn that were discovered from the Earth. Voyager spacecraft photography reveals it to have a crater so large that if it had been struck by a somewhat larger comet, it would have shattered. Comets can destroy worlds. (Painting by Kim Poor.)*

Chapter 15

THE WRATH OF HEAVEN:
I – THE GREAT DYING

A comet never descends all the way to the lowest regions of the atmosphere, and does not approach the ground.

Seneca,
Natural Questions,
Book 7, 'Comets'.

Comets are such beings as have been on account of their merits raised to heaven, whose period of dwelling in heaven has elapsed and who are then redescending to Earth.
Al Biruni.

The ground is growing. Silt settles onto the stream bed, dust falls from the air, and the ground is a little higher. A pretty typical number is a micron a year – a film at most one microscopic particle thick, an almost imperceptible annual patina settling over the Earth's surface. A micron is almost nothing. But over geological time, it adds up. In 10,000 years, at this rate, you build up a centimetre of the stuff, and in a million years, a metre. If you could find a place where, over a good part of the age of the Earth, material was only added and never removed, you might discover cliffs kilometres high in which an undisturbed record of the history of our planet is preserved.

Providentially, there are such places; one of the best known is the Grand Canyon of North America, where the sedimentary column has been exposed by the erosion of the Colorado River. You see a lovely layering in the rocks, a sequence of muted pastels. In adjacent cliff faces, there are almost identical patterns. Each layer is an epoch. Every boundary between layers corresponds to some significant change in the environment. If the sedimentary columns show the same pattern everywhere, the environmental changes must have been worldwide. As you glance from the top to the bottom of such a cliff, you see the history of the world. You have only to step up and read.

Without doing any digging at all, you might see something bright gleaming in the sunlight. You examine it more closely and realize it is a fossil – perhaps a dinosaur's knee, or the jaw

15.1 *(Opposite) A large cometary nucleus, ten kilometres across, is just about to strike the Earth – 65 million years ago in this schematic representation. The resulting explosion is thought to have propelled huge clouds of pulverized ocean bottom all over the Earth, producing a time of cold and dark which resulted in the extinction of the dinosaurs and many other species. The Nuclear Winter is a similar effect. (Painting by Don Davis.)*

15.2 *An ammonite from the Cretaceous seas. Because they made abundant fossils the shapes of ammonite shells are known very well. (Drawing by Maren Leyla Cooke.)*

of an armoured fish. With a microscope, you easily discover fossils of little creatures, perhaps single-celled plankton that floated in some primeval sea. You look elsewhere, at other sedimentary columns, and the remarkable thing is that everything goes together. This particular thick, reddish deposit tends, over much of the world, to be underneath these two thin grey layers; it always has a similar mix of fossils. You find an ammonite shell, and it confirms that the vast desert landscape of which this cliff is part was once the bottom of an ocean. A few metres above or below, and the fossils look very different; you realize that if you wait long enough, the landscape will change dramatically.

The rocks are trying to tell us something. These fossils are all that remain of creatures that once strode, or slithered, or swam, or just took root, beings that arose and died out before there were any humans to take note of them. You take a particular fossil and, all over the world, there is a layer – corresponding to some epoch in the past – below which that organism had not yet come into being. Generally, there is a layer above, in which that fossil no longer appears, while the fossils of new organisms are found for the first time. The lesson is clear: great numbers of species have become extinct, providing opportunities for other species to evolve and fill the vacated ecological niches. You count up all the beings that leave fossils – many were too small to be seen without a microscope, although some were as big as an office building – and you discover that the vast majority by far of all the species that ever lived are now extinct. Extinction is the rule. Life is the exception.

And then you notice that the different fossil forms do not succeed each other at a uniform pace. Instead, the rocks proclaim long periods in which life changed very little, punctuated by short intervals of cataclysm, turnover, the wholesale loss of many kinds of creature all over the planet at once – followed astonishingly quickly by many new forms, clearly evolved from the survivors of the catastrophe. The clearest boundaries in the sedimentary column are the ready markers of fearsome, planet-wide disasters. You have a planet where, for tens of millions of years, everything is fine. It is enough to give even inveterate pessimists a sense of security. But then, maybe when you'd least expected it, tumultuous changes occurred worldwide – changes so striking that they are instantly apparent to the eye and mind of observers like us, who have evolved into existence hundreds of millions of years later.

There is a steady background rate of extinction, of course. In our time, because of human activities, species become extinct every year – from land reclamation or industrial pollution, or

15.3 *A Labyrinthodont, an amphibian from the late Triassic period. (Drawing by Maren Leyla Cooke.)*

cutting down the forests. The passenger pigeon and the great auk became extinct in the nineteenth century. The last mastodon died as recently as 2000 BC, the giant armadillo around 5800 BC, and the giant ground sloth (it browsed on treetops) around 6500 BC. We may have been responsible for the demise of many of these creatures. Such extinctions are part of the natural ebb and flow of life, and are caused by minor changes in the physical or biological environment. What we are concerned with here is something quite different: massive extinctions of dozens or hundreds of families of life simultaneously, all over the Earth.

The extinction of what biologists call a 'family' of plants or animals implies a major loss. Think of the variety of dogs in the world, from chihuahuas to great danes. They are all one species because they are all interfertile. Biologists then distinguish a broader category, called a genus, that embraces not only dogs, but wolves and jackals as well. This genus, *Canis,* is in turn part of a much larger category, a family that includes foxes and hyenas; together they are all the *Canidae.* A family is a major group of beings. The human family, the family of the hominids, includes most of those primates who walked around on two feet and tried to puzzle things out over the last few million years, even though we would not recognize many of them as human were we to meet

15.4 *The Brontosaurus, a dinosaur as big as a house that flourished in North America in the late Jurassic. The Brontosaurus was made extinct long before the Cretaceous period, and is an example of 'background' extinctions: species die out at a steady pace over the whole history of the Earth, on top of which there are short periods of mass extinction. (Drawing by Maren Leyla Cooke.)*

SOME MASS EXTINCTIONS

Geological period in which extinctions occur	Millions of years ago	Percentage of marine families extinct*
Ordovician	435	>25%
Devonian	365	>25%
Permian	245	>55%
Triassic	220	>40%
Cretaceous	65	20%
Eocene	35	15%

*Rough estimate of percentage of all families of marine animals with hard parts (so we have fossil evidence of their existence) rendered extinct; numbers are given to nearest 5 million years (column 2), or nearest five percent (column 3); > means 'more than'. Extinctions of land families are thought to be roughly comparable. Data courtesy J. John Sepkoski Jr, University of Chicago. As described in Chapter 16, there is active debate about some of these numbers. The Permian and Cretaceous mass extinctions are the most reliably determined.

15.5 *A trilobite from the Cambrian period. The last trilobites died out in the Permian period. (Drawing by Maren Leyla Cooke.)*

15.6 *An armoured fish, none of which survived the Devonian period. (Drawing by Maren Leyla Cooke.)*

15.7 *One of the first ocean corals, the rugose coral, gone with the Permian extinctions. (Drawing by Maren Leyla Cooke.)*

15.8 *Dimetrodon, a sail-finned dinosaur, one of the casualties of the Cretaceous extinctions. (Drawing by Maren Leyla Cooke.)*

them on the street. The loss of a family represents the hacking off of a branch of the tree of life.

In the table on p.237, six of the most spectacular known interruptions in the history of the Earth are listed, along with an indication of how massive these extinctions were. It is with a sense of shock that we look at the last column in the table. It shows, for each of the six catastrophes, the percentage of all the families of known marine organisms that were lost. At the Ordovician and Devonian boundaries, fully a quarter of the families became extinct, less for the Eocene event, more for the Triassic. In the Permian catastrophe, more than half of all the families of life on Earth were lost, and more than ninety percent of all species. One of the most recent of these planetary disasters is at the end of the Cretaceous period, some 65 million years ago, when something like twenty percent of the families and seventy-five percent of the species disappeared.

Some of the creatures that were lost in these extinctions are pictured on these pages – a brachiopod that attached itself to the ocean bottom, extinct by the Ordovician; a fish in an armoured shell, extinct at the end of the Devonian; a trilobite – they hunted in herds on the ocean bottoms and had existed for almost 300 million years – extinct in the Permian; a kind of coral, also gone in the Permian; an ammonite, a relative of the octopus, extinct in the Cretaceous; and triceratops, one of the last of the dinosaurs, with three very nasty horns and clanking bone armour, also a Cretaceous casualty. You take one look at triceratops, and you can tell that times were dicey in more ways than one towards the end of the Cretaceous Period.

Scientists look closely at the Cretaceous extinctions because they concern us particularly. A critical turning point in the evolutionary path to humans can be glimpsed among the reptile bones. The Cretaceous catastrophe wiped out every family, every genus, and every species of dinosaur, and they were as varied and successful as the mammals are today. It is as if all the mammals one day were to roll over, their appendages stiffly in the air – every shrew and whale, and every person. All the flying and swimming reptiles died as well, and more than a hundred families of beings that live in the oceans. It was a catastrophe enormously beyond anything humans have ever known, at least so far.

The first mammals arose at about the same time as the first dinosaurs. The dinosaurs were the lords of the Earth, the largest creatures, the most powerful, the ones who would capture your attention on any Cretaceous landscape. The mammals, our ancestors, were then tiny, furtive, scampering creatures – cautious, mouse-like, they spent much of their time keeping out of the way of the thundering reptiles. A few dozen little mammals would have made a meagre lunch for a middling-sized carnivorous dinosaur. For over 100 million years, our ancestors were at an apparent evolutionary dead end, living at the margins of a world dominated by dinosaurs. If you surveyed that late Cretaceous landscape – in which the trees and flowers looked pretty much as they do today, but where the dominant animals were all reptiles – you would not have bet much on the chance of our ancestors.

15.9 *Triceratops, the last and largest of the horned dinosaurs, extinct at the end of the Cretaceous period. It was bigger than an elephant. (Drawing by Maren Leyla Cooke.)*

But the dinosaurs went and died, every last one of them, and left the planet to the mammals. Mammals were not the only ones to survive, of course: snakes and salamanders were in good shape, as were fish and insects and crocodiles, and many land plants. But the mammals soon predominated. Tentatively at first, and then with an increasing exuberance, the mammals evolved, grew, diversified, taking advantage of the demise of the competition. We owe our very existence – every one of us bigger than a mouse, anyway – to the extinction of the dinosaurs. So perhaps we may be excused for having a special interest in this question: why did the dinosaurs, along with most of the other species on Earth, abruptly die some 65 million years ago?

15.10 *A sessile shelled animal, from the late Ordovician period. (Drawing by Maren Leyla Cooke.)*

In Part I of Laplace's *System of the World*, there is a chapter entitled 'On Comets and Superstition', in which, like many others, the great French scientist at once criticized as superstition the worldwide human fear of comets, and advanced as sober scientific hypothesis a good and practical reason for dreading them:

> In those times of ignorance, mankind were far from thinking that the only mode of questioning nature is by calculation and observation: According as phenomena succeeded with regularity or without apparent order, they were supposed to depend either on final causes or on chance; and whenever any happened which seemed out of the natural order, they were considered as so many signs of the wrath of heaven.
>
> But these imaginary causes have successively given way to the progress of knowledge, and will totally disappear in the presence of sound philosophy, which sees nothing in them, but expressions of the ignorance of the truth.
>
> To the terrors which the apparition of comets then inspired succeeded the fear, that of the great number which traverse the planetary system in all directions, one of them might overturn the earth.

Laplace himself, we remember, had carefully studied the

orbital evolution responsible for the near impact with the Earth of Comet Lexell in 1770 (see Chapter 5):

> They pass so rapidly by us, that the effects of their [gravitational] attraction are not to be apprehended. It is only by striking the earth that they can produce any disastrous effect. But this circumstance, though possible, is so little probable in the course of a century, and it would require such an extraordinary combination of circumstances for two bodies, so small in comparison with the immense space they move in, to strike each other, that no reasonable apprehension can be entertained of such an event.

So, is the fear of a cometary collision with the Earth in the same category as superstitious dread of comets? Not at all:

> Nevertheless, the small probability of this circumstance may, by accumulating during a long succession of ages, become very great. It is easy to represent the effect of such a shock upon the Earth: the axis and motions of rotation changed, the waters abandoning their previous position, to precipitate themselves towards the new equator; the greater part of men and animals drowned in a universal deluge, or destroyed by the violence of the shock given to the terrestrial globe; whole species destroyed; all the monuments of human industry reversed; such are the disasters which a shock of a comet would produce.

The expectation that comets – through universal deluge or other means – are in fact harbingers of disaster has been a respectable speculation for the whole scientific history of the subject, dating back to Edmond Halley himself. The brand of mischief that comets are said to bring – flood, darkness, fire, rending the Earth asunder – changes with time and astronomical fashion. But the association of comets with catastrophe remains steady, although few consider this association in modern times to be a retroactive justification of superstitious excesses in generations past.

Laplace connected this awful vision of a comet striking the Earth with an apparent paradox in time-scales. Human history was only a few thousand years old. But Laplace knew – from arguments such as those of Halley on the salt content of seawater – that the Earth was much older. Laplace, the cosmic evolutionist, had not a glimmering of biological evolution. Charles Darwin's *The Origin of Species* was sixty years in the future. Laplace could not imagine the world existing for long periods before there were humans. So he had to ask himself why human history and human civilization were not much older?

> We see ... then why the ocean has abandoned the highest mountains, on which it has left incontestable marks of its former abode: we see why the animals and plants of the south may have existed in the climates of the north, where

their relics and impressions are still to be found: lastly, it explains the short period of the existence of the moral world, whose earliest monuments do not go much farther back than 3,000 years. The human race reduced to a small number of individuals, in the most deplorable state, occupied only with the immediate care for their subsistence, must necessarily have lost the remembrance of all sciences and of every art; and when the progress of civilization has again created new wants, every thing was to be done again, as if mankind had been just placed upon the earth.

There is here not only the explicit suggestion of global cometary disasters, but, also, cometary extinctions; and even a hint that such catastrophes occur episodically, through the whole previous history of the Earth. Laplace was two centuries ahead of his time.* After Laplace, the association of comets with catastrophe was popular. Some authors imagined debris from the comets – the clay in Donnelly's *Ragnarok*, for example – distributed over much of the Earth; others imagined the effects of the collision to be felt only in a limited area. Occasionally, consequences were imagined that were even more dire than those Laplace had depicted. In 1893, the French writer Camille Flammarion – who seemed to have had a penchant, even beyond that supported by the facts, for frightening people with comets – wrote a science fiction story called *The End of the World*:

> Like a great celestial projectile the solid nucleus of the comet pierced the egg-shell crust of the Earth and buried itself in the semi-molten interior. The comet tore its way on like a shot piercing the boiler of a battleship. The Earth was immediately converted into a planetary volcano. Oceans were spilled like thimbles of water ... continents were twisted and torn like paper.

This is rather overdoing it. But if comets are kilometres across and travelling at great speed in the same part of the solar system as the Earth, it follows that, sooner or later, a largish comet will indeed strike the Earth, with consequences that we can be sure would be catastrophic, even if we were unable to trace exactly what would happen. This is the basis of Laplace's argument. It is easy to calculate – it can be done on the back of the proverbial envelope – roughly how long the Earth waits between successive collisions with sizeable comets. In the first decade of the twentieth century, W. H. Pickering of Harvard calculated that a fair-sized cometary nucleus should strike the

*She first beholds the raging comet rise,
Knows whom it threatens, and what lands
destroys.*

Juvenal,
The Sixth Satire
(translated by John Dryden, 1693).

---

*Nevertheless, the advance of scientific knowledge has overtaken many of Laplace's arguments. We know that humans were on Earth for a million years before we invented written history and 'civilization'. The fossil evidence that some mountainous terrain was once beneath the oceans puzzled many others, including Petrarch and Leonardo da Vinci. The explanation, firmly established by modern geology, is not of a universal deluge which covered the mountaintops, but that some mountaintops have slowly raised themselves from the ocean floor.

Earth once every forty million years. But the dozens of collisions he imagined the Earth to have suffered during its history seemed to have done little damage, a conclusion he deduced from the fact that life is all around us.

The leading astronomy textbook in America after the end of the Second World War* included this paragraph:

> It is probable that the Earth has undergone many collisions with comets during geological time. It may readily be computed that a small, rapidly-moving body which aproaches the Sun within one Astronomical Unit stands about one chance in four hundred million of hitting the earth. As about five comets come within this distance every year, the nucleus of a comet should hit the earth, on the average, once in about eighty million years.

In the light of recent work, this seems almost prescient. But because the authors had accepted the sand bank hypothesis of comets (see Chapter 5), they disbelieved in massive cometary nuclei. So they explicitly stated that collisions of comets with the Earth 'would probably fall very short of producing wholesale destruction of terrestrial life'. Throughout the twentieth century, scientists intermittently connected biological extinction with cometary impacts, although the prevailing wisdom held that a catastrophe on such a scale could not be visited by so minor a body.

About halfway from Florence to Rome on the *autostrada*, you make a left turn, drive through Perugia, and continue an equal distance up into the Apennines. Eventually, you come to a little village called Gubbio; it dates back to medieval times. There, at the side of the road, is the sequential banding of a wonderfully preserved sedimentary column; it dates back to much more ancient times. If you go up to it, you can see a thin layer of pink and grey sitting above a bright white bed of rock. It marks the end of the Cretaceous Period.

You cut out a little piece and bring it back to the laboratory. The white rock is limestone, and under the microscope you can see a matrix of calcite plates and shells manufactured by the microscopic plants and animals that lived in the warm seas. The chalk of the white cliffs of Dover, in Britain, were manufactured by these calcite-secreting oceanic micro-organisms, which became extinct in the Cretaceous disaster. In fact, this is where the Cretaceous gets its name: *creta* is Latin for chalk. At Gubbio, these oceanic limestones are punctuated by the grey/pink layer; it is about a centimetre thick, and made of clay. The clay and the limestone both must have accumulated as individual grains, and settled through the

---

Astronomy. I. The Solar System by H.N. Russell, R.S. Dugan and J.Q. Stewart (revised edition, Boston, Ginn and Co, 1945).

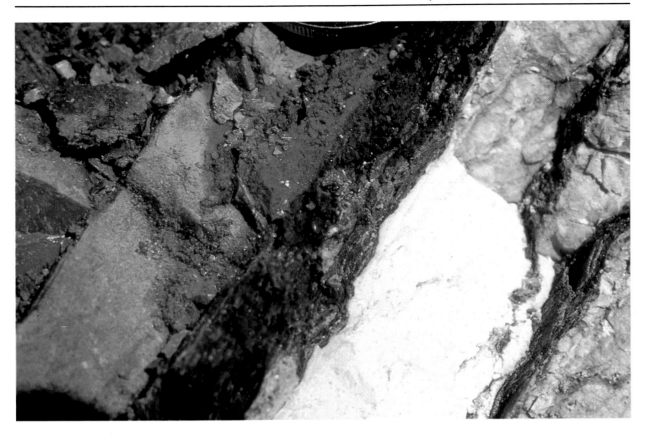

quiet waters of the deep ocean tens of millions of years ago. But, just above the clay, the fossils in the limestone are very different from those just below. You can tell at a glance that the clay marks a catastrophe.

Similar layers are found at the Cretaceous Boundary all over the world. The kind of microscopic pollen from flowering plants changes abruptly above the clay layer compared to below it – the land equivalent of the oceanic extinctions of micro-organisms. Below the Cretaceous clay, you find the fossils of dinosaurs who prowled the Earth for 160 million years or more. Above it is the time of the mammals.* This grey boundary marks a catastrophe that ended the Cretaceous Period and much of life on Earth. What is in that clay? Are there any clues to what caused the Great Dying?

Gold and platinum are valued in large measure because they are rare. But when you look at the spectrum of the Sun and the stars, or examine a meteorite newly fallen to Earth, you find that there is relatively much more of these precious metals up in the sky. Not that meteorites have lodes of gold or veins of platinum, but, compared to some plentiful element such as silicon, the Earth seems anomalous, strangely

15.11 *The record in the Gubbio rocks. This is the sharp boundary in the roadside rocks at Gubbio, Italy, of the end of the Cretaceous period and the Mesozoic Age of geological time. The Cretaceous rocks (bottom right), are clearly shown to be white limestone when broken up a little with a hammer. The colour of this chalk layer is due to the fossils of micro-organisms that lived in the late Cretaceous oceans. The reddish brown rocks (top left) are from the Tertiary period, in which no dinosaur fossils can be found and where mammals are in the ascendant. Snaking through this picture along the diagonal, separating the Cretaceous from the Tertiary rocks, is a layer of grey clay. It corresponds exactly to the time that the last dinosaurs appear in the fossil record, and is rich in iridium. (Courtesy, Walter Alvarez, University of California at Berkeley.)*

*As for humans, we have been here for one percent of the time the dinosaurs ruled. There is no trace of us except at the very top of the sedimentary column.

depleted in precious metals. Now in molten rock, gold and platinum tend to move with iron. And the iron that was once uniformly mixed through the bodies that formed the Earth is now mainly concentrated in the liquid core of our planet, 3,000 kilometres beneath our feet. It is a good bet that most of the gold and platinum migrated there, along with the iron, when the newly-formed Earth was molten. This is true of other elements as well, elements less well known than gold and platinum – in particular, the elements iridium, osmium, and rhodium. So if iridium were present in sufficient quantities in some terrestrial sediment, it might be a tell-tale signature of extraterrestrial intervention in earthly affairs.

In the late 1970s, a group of scientists at the University of California at Berkeley began to wonder about the clay at the Cretaceous Boundary. The most prominent members of the team were Luis Alvarez, a Nobel Laureate nuclear physicist, and his son, Walter Alvarez, a geologist, both professors at Berkeley. The elder Alvarez proposed using a technique called neutron activation analysis, which could measure extremely small quantities of, among other things, iridium. The Alvarez team had the presence of mind to examine the iridium content in, above, and below the clay layer at Gubbio that marks the end of the Cretaceous. They measured the abundances of twenty-eight chemical elements through this portion of the sedimentary column, and found something astonishing.

THE WEDDING RING ANOMALY

Of the dozens of iridium concentrations that have been found at the Cretaceous Boundary, one turns out to be spurious. In ordinary rock from the crust of the Earth, the iridium content is less than a tenth of a part per billion. Yet even this is enough to be measured by neutron activation analysis. The abundance at Gubbio was much more – six parts per billion – and elsewhere still larger. But the iridium in one laboratory specimen of Montana Cretaceous clay turned out to be:

> due to the platinum wedding or engagement ring worn by a technician who had prepared the samples for analysis. Platinum used for jewellery contains about ten percent iridium... If a platinum ring loses ten percent of its mass in 30 years, the average loss per minute, if it all deposits on a sample, is about [a hundred times] higher than our sensitivity of measurement.*

The Alvarezes and their colleagues conclude that a few seconds exposure to a platinum wedding ring minding its own business is enough to produce a spurious signal in the analysis. The more sensitive the instrument, the more careful you must be. Technicians, henceforth, must wear gloves.

*Walter Alvarez, Frank Asaro, Helen V. Michel, Luis W. Alvarez, 'Iridium Anomaly Approximately Synchronous with Terminal Eocene Extinctions' Science, 216, p. 886 (1982).

Twenty-seven of the elements showed no major changes in abundance in and out of the layer. But iridium was enriched by thirty times in the clay. Similar results have now been obtained from all over the world. In Haiti, there is roughly 300 times more iridium at the Cretaceous Boundary than in adjacent layers; in New Zealand 120 times more; on the shores of the Caspian Sea seventy times; in Texas, forty-three times; and in the deep ocean of the the northern Pacific, 330 times.

A worldwide iridium-rich layer marking the end of the Cretaceous Period looks very much like direct evidence that a large cosmic body struck the Earth 65 million years ago. You can even calculate how big the body had to be to distribute this much iridium over the Earth. The answer turns out to be about ten kilometres across, a fairly typical size for a cometary nucleus or an asteroid. Of the four comets whose diameters were measured by radar techniques through the middle 1980s, two of them were of this magnitude. Many Earth-crossing asteroids seem to be extinct comets (see Chapter 14). The chances seem to be better than even that the Cretaceous catastrophe was triggered by a comet that hit the Earth.* If so, that layer of clay is comet stuff – mainly minerals, because the organics should long ago have been eaten by the microbes of Earth.

The iridium was not deposited instantaneously, but over 10,000 or even 100,000 years – much longer than the timescale of a single impact. But (see Chapter 16) many comets may have been involved, and particles excavated by a single impact may have continued to fall back to Earth for long periods of time. Occasionally, volcanoes can produce anomalously high concentrations of iridium, but material from the Cretaceous Boundary shows alterations in the form and chemistry of minerals that are consistent only with an enormous shock – which can be provided by a cometary impact, but not by a volcanic eruption.

We have heard something like this before – catastrophes caused by a comet coming close to the Earth and raining down a worldwide layer of clay – the very thesis propounded by Ignatius Donnelly in *Ragnarok* (see Chapter 10). The catastrophes are different, the timescale is different, but the idea is unmistakably similar. The clay layers of Donnelly were not at the Cretaceous Boundary, Donnelly knew nothing of iridium, and the Alvarezes were not inspired by Donnelly. The iridium speaks for itself.

If an object ten kilometres across hit the Earth at cometary velocities, it woud carve out a huge crater, more than a

It does seem possible and even probable that a comet collision with the Earth destroyed the dinosaurs and initiated the Tertiary division of geologic time... It will most probably be millions of years before the next collision occurs.

Harold C. Urey,
'Cometary Collisions and
Geological Periods'
Nature, 242, page 32 (1973).

The star which rules thy destiny
Was ruled, ere earth began, by me:
It was a world as fresh and fair
As e'er revolved round sun in air;
Its course was free and regular,
Space bosom'd not a lovelier star.
The hour arrived – and it became
A wandering mass of shapeless flame,
A pathless comet, and a curse,
The menace of the universe...

What wouldst thou, Child of Clay, with me?...

The comets herald through the crackling skies;
And planets turn to ashes at his wrath.

Lord Byron,
Manfred I, i, 118 and II,
iv, 381; (1816).

*While in the following pages we do not everywhere add the words 'or asteroid' after 'comet', we stress that there is a significant probability that the Cretaceous impact was caused by an asteroid not of cometary origin.

hundred kilometres in diameter. This would be true whether the comet struck on land or ocean, because the depth of the ocean is considerably less than the size of the comet (see Figure 8 in Chapter 7). The resulting debris – some mix of pulverized comet and pulverized Earth – would be thrown high; indeed, much of it would be ejected well above the atmosphere and out into space. The debris, therefore, should have been transported all over the Earth, as is observed. The clouds of fine particles, ejected to escape velocity, would travel around the Sun in orbits that repeatedly intersected the orbit of the Earth. It is possible, therefore, that a steady rain of fine debris would have fallen for tens of thousands of years, until the cleansing of the inner solar system by the Poynting-Robertson Effect (see Chapter 14) had been completed.

A crater a hundred kilometres across is too prominent to miss. Where is it? Only three are known. Two of them are much too old, having been formed more than 600 million years ago. The remaining one, Popigay Crater in Siberia, seems to be some 30 million years old, and is therefore too young. But the Earth is two-thirds ocean, and the Cretaceous crater was most likely formed in the ocean depths. Such a crater could be almost anywhere in the deep sea today without anyone being the wiser. There is, in the unclassified literature at least, no detailed topographical mapping of the deep-sea floor. Perhaps

Topographical maps of three neighbouring worlds – Mars 15.12 (below), Earth 15.13 (opposite above), and Venus 15.14 (opposite below). On Mars, there are thousands of impact craters too small to appear at this resolution, but very large craters as well – the two largest, shown here, are Hellas (near 300° longitude) and Argyre, both at southern mid-latitudes. In the absence of oceans or a thick atmosphere, Mars preserves large impact craters from a billion years ago or more. The Earth, on the other hand, has such efficient erosion due to the atmosphere and liquid surface water, as well as plate tectonics, that there are relatively few impact craters preserved, and only a handful dating back to the early history of the planet. Venus, with a dense atmosphere and a high surface temperature (although no recent oceans), seems closer to the Earth than to Mars in its retention of large, ancient impact scars. These maps, based partly on spacecraft data, were prepared by Michael Kobric of the Jet Propulsion Laboratory, NASA and US Geological Survey.

the evidence of a vast crater remnant sits in some Soviet or American hydrographic office, available only on a need-to-know basis to the officers who command nuclear submarines. But, worse than that, we know that the Cretaceous ocean bottoms have largely been destroyed over the intervening ages by the movement of continental-sized plates back into the Earth's interior. Somewhere on the Earth, there was once a crater the size of Belgium, or Corsica, or Swaziland. Like the smallpox virus, the comet left a small disfigurement as a reminder of how a near-fatal disease was survived. Now, the pockmark may be gone, and the thin layer of clay that fell from the sky may be our sole souvenir, apart from the break in the fossil record, of the great Cretaceous catastrophe.

But how can a cometary impact kill hundred-ton land-dwelling dinosaurs as well as microscopic algae in an ocean halfway around the world? A number of suggestions have been made: perhaps cometary cyanides (see Chapter 8) poisoned everybody, or toxic metals, or acid rain. But the probable prime mechanism was suggested by the Alvarezes themselves: if you lift a layer of fine particles one centimetre thick off the Earth and up into the stratosphere, the individual particles will take years to fall out. And since a centimetre thickness of clay covering all the Earth is opaque, it must also

15.15 *A cometary impact with the Earth in the late Cretaceous period raises an immense cloud of fine particles that are blown by the prevailing winds. No familiar continent can be made out; because of plate tectonics, the continents of the Earth were very different 65 million years ago. Omitted (for clarity) in this painting is the large amount of impact debris that would immediately be thrown high into the atmosphere and fill the air all over the planet. By blocking sunlight, this global obscuration is thought to have cooled and darkened the Earth. (Painting by Jon Lomberg.)*

be opaque when it is distributed throughout the upper atmosphere, and was slowly falling back down. Sunlight would not have penetrated this cloud of iridium-rich cometary clay. For months, or even years, the Earth would have been darkened and cooled.

PREMONITION

The consequences of a collision with [a small earth-crossing asteroid] are unimaginable; the repercussions would be felt the world over. In dissipating the energy equivalent of half-a-trillion tons of TNT, 100 million tons of the earth's crust would be thrust into the atmosphere and would pollute the earth's environment for years to come. A crater fifteen miles in diameter and perhaps three to five miles deep would mark the impact point, while shock waves, pressure changes, and thermal disturbances would cause earthquakes, hurricanes, and heat waves of incalculable magnitude. Should [the asteroid] plunge into the ocean a thousand miles east of Bermuda, for example, the resulting tidal wave, propagating at 400 to 500 miles per hour, would wash away the resort islands, swamp most of Florida, and lash Boston – 1500 miles away – with a 200-foot wall of water... The energy involved is the equivalent of 500,000 megatons of TNT – two orders of magnitude above that involved in the largest recorded earthquake, and four or five orders of magnitude more than Krakatoa... If the strike occurred in midocean, tsunamis in the 100-foot category would cause worldwide damage. If the strike occurred on land, the blast wave would level trees and buildings within a radius of several hundred miles, and some 10^8 tons of soil and rock dust would be thrown into the stratosphere, where for several decades it would act to reduce the solar radiation ordinarily received at earth's surface and threaten the triggering of an ice age.

M.I.T. Student Project in
Systems Engineering,
'Project Icarus' (MIT Press,
Cambridge, Massachusetts, 1968).

This paragraph represents an important premonition of the Cretaceous impact, and of nuclear winter.

The average temperature of the Earth during the Cretaceous Period was some ten degrees centigrade warmer than it is today; by current standards, the Earth was then a tropical planet, with many life forms unprepared for bitter cold. Plants and animals in the tropics today, where the temperature never falls below freezing, have few defences against a major freeze. Calculations show that, immediately after the Cretaceous impact, it would have become very cold all over the Earth – perhaps tens of degrees below freezing. Also, for months, the amount of light that reached the Earth's surface would have been too low for most plants to photosynthesize, and even too low for animals to see. Through seeds, spores and the like,

15.16 *A Triceratops wanders forlornly over the frozen and darkened late Cretaceous landscape in this illustration of the probable consequences of a cometary impact 65 million years ago. (Painting by Don Davis.)*

land plants might have survived many years of cold and dark. But microscopic plants in the oceans, with no food reserves, would have died quickly – and the entire oceanic food chain which depends on those plankton would have collapsed soon after. The plight of the dinosaurs, stumbling across a frozen, sombre, devastated landscape, unable to find either food or warmth, can readily be imagined.

The Cretaceous extinctions thus seem to have a cause very similar to the consequence of nuclear war which is known as 'Nuclear Winter'. Through the dust excavated by nuclear groundbursts, and the smoke from the burning of 'strategic targets' in and around cities, we humans, if we are not very careful, can generate our own climatic catastrophe, perhaps adequate to bring about extinctions in our age as massive as those of the Cretaceous Period. The principal difference would be that the dinosaurs were innocent.

It seems likely that had not a comet or asteroid hit the Earth 65 million years ago (or subsequently), the dinosaurs would still be here, and we would not. We would be merely one of a countless number of unrealized possibilities in the genes and chromosomes of the other beings of Earth – as we can be again if we are not very careful.

Chapter 16

THE WRATH OF HEAVEN:
II – A MODERN MYTH?

Believing that every object and every event in the universe is arranged and directed by an Omnipotent Contriver, we must admit that when the Almighty formed the wondrous plan of creation, 'foreseeing the end from the beginning', he arranged the periods and the velocities of comets in such a manner that, although occasionally crossing the planetary orbits, they should not pass these orbits at the time when the planets were in their immediate vicinity. And should such an event ever occur, we may rest assured that it is in perfect accordance with the plan and the will of Omnipotence, and that it is, on the whole, subservient to the happiness and order of the intelligent universe, and the ends intended by the Divine government.

> Thomas Dick, *The Sidereal Heavens*
> *and Other Subjects Connected*
> *With Astronomy, As Illustrative*
> *of the Character of the Deity*
> *and of an Infinity of Worlds*
> (Philadelphia, 1850).

I feel rather at a disadvantage in speaking of comets to you; comets nowadays are not what they used to be.

> Arthur Stanley Eddington,
> 'Some Recent Results of
> Astronomical Research',
> Friday Evening Discourse,
> The Royal Institution,
> (London, 26 March 1909)

Stated baldly, it still sounds like something from the early pulp days of science fiction: 65 million years ago, a comet came out of space, hit the Earth, and extinguished much of the life on this planet. The Alvarez discovery has created something like a scientific revolution, with consequences spreading through geology, astronomy, and evolutionary biology, and even making incursions into international politics and nuclear strategy. The connection between cosmic events and our own existence is stirring.

But the Cretaceous extinctions are not the only, or even the major, instance of mass extinctions in the history of life on Earth. And – as Laplace, and many others, realized – there must have been many impacts of comets (and asteroids) with

16.1 *(Opposite) The dust lanes in the plane of the Milky Way galaxy and its bulging core are shown in the distance, as the Sun and the Earth approach a giant interstellar cloud. (Painting by Jon Lomberg.)*

the Earth. Iridium concentrations have been found in another sedimentary layer, associated with another mass extinction (although Permian sediments – associated with a major mass extinction on everybody's list – shows so far not a hint of an iridium anomaly). There are two layers of microtektites (see box, page 256) near the boundary of the Eocene extinctions, suggesting at least two separate impact events then which spattered droplets of molten rock over the world. The Alvarez discovery set many other scientists, from a range of disciplines, thinking about cometary impacts and mass extinctions, which has led to still more extraordinary claims. However, these recent developments are more speculative, and – since some of the ideas contradict each other, and for other reasons – we should here tread cautiously.

The new developments began at the University of Chicago, where the American paleontologist J. John Sepkoski had been compiling a master list of all the families of life that have become extinct in all the sedimentary rocks that have been studied. Sepkoski tabulated the epoch in which each extinct

16.2 *A dinosaur contemplates its fate as the sky is ablaze with comets during the postulated late Cretaceous comet shower. The dinosaur depicted had something like hands, and a larger brain size for its body-weight than most of its contemporaries. If the dinosaurs had not been rendered extinct, the dominant form of intelligent life on Earth today would perhaps have descended from such a creature. (Painting by Jon Lomberg.)*

family died out. He could see that the background rate of extinction, more or less constant through geological time, was due to many causes – mountain building, greenhouse effects, disease, Darwinian competition, and the like. Superimposed on this background were times when the extinctions all piled up, one on top of another, as at the end of the Cretaceous Period, and during the other Great Dyings (see table in Chapter 15). So far, nothing much new – although the family extinction data were encyclopaedic.

But when Sepkoski and David Raup, also of the University of Chicago, analyzed the extinction dates, they were surprised to find what looked like a periodicity. Every 30 million years or so, like clockwork, it seemed, plants and animals drop dead all over the planet. At the family and species level, a major fraction of all life on Earth becomes extinct at these apparently regular intervals, they concluded. Scientists tend to talk about it with detachment, but there is something unnerving about these countless lives of other beings, our cousins, being snuffed out, their ancestral lines rendered meaningless, by some regular motion of the Grim Reaper's scythe. Naturally, we wonder when our time will come.

Paleontologists have for decades invoked many different explanations of the mass extinctions, but none of the proposed mechanisms had a period built in. The Earth does not seem to have any 30-million-year-long cycles – not in volcanoes, not in plate tectonics, not in the weather. Periods of this length belong in the domain of astronomy.

The iridium in the Cretaceous Boundary has made the idea of periodic (or at least episodic) mass extinctions, visited on the Earth from space, more palatable; but for many scientists, and others, it remains a bitter pill to swallow. Even if, for the moment, we grant that small worlds fall out of the sky like clockwork, shouldn't this fact be evident from the ages of craters on Earth? The craters are there – lots of little recent ones not yet eroded away, and smaller numbers of larger and older craters, the biggest being more than 100 kilometres across. Still larger craters must have been made, of course, but that was so long ago in the Earth's history that the slowly crashing and diving plates that make up the Earth's crust would have destroyed them long ago. Thus newly-motivated, several teams of scientists have examined the ages of the surviving craters, and have seemingly found – to simultaneous satisfaction and amazement – that craters tend to have been excavated, by the impact of objects from space, roughly every 30 million years. What's more, the events are reported to have been in phase: mass extinctions tend to happen when large craters are formed, the two events presumably caused by the same impacting bodies.

The implications of periodicity for evolutionary biology are profound. The most obvious is that the evolutionary system is not 'alone' …With kill rates for species estimated to have been as high as 77% and 96% for the largest extinctions, the biosphere is forced through narrow bottlenecks and the recovery from these events is usually accompanied by fundamental changes in biotic composition. Without these perturbations, the general course of macroevolution could have been very different.

David M. Raup and
J. John Sepkoski, Jr,
'Periodicity of Extinctions
in the Geologic Past,
*Proceedings of the National
Academy of Sciences of the
USA,* Volume 81, page 801,
(1984).

TEKTITES

Certain sedimentary layers of the Earth are strewn with glassy, geometrically-smooth inclusions, ranging from a few centimetres in size to sub-microscopic. That tektites are produced when comets hit the Earth was first proposed by the American chemist Harold Urey in 1957, and – while not definitively decided – the hypothesis has stood the test of time. The comet strikes the Earth, digs out a large crater, and melts the underlying ground in the process; droplets of silicates are flung over huge distances, freezing in the process and producing streamlined, sometimes teardrop-shaped, forms. The tektites themselves show signs that they were cratered during these violent events. Those shown here were produced a little over 35 million years ago. It has been claimed, and disputed, that layers of sediment containing the so-called 'microtektite horizon' are associated with the Eocene extinctions. One of the leading experts on these tiny, glass forms is named, appropriately enough, Billy Glass. He is at the University of Delaware.

16.3 *Three sample of microtektites: right, from Australasia, at a magnification of fifty times; opposite above, from the Ivory Coast at a magnification of 190 times, and opposite below, a microtektite about 240 microns in diameter with strange star-shaped craters. (Courtesy, B.P. Glass, University of Delaware.)*

The correlation is not one-to-one. In sediments older than the Cretaceous, dating of these boundaries has an uncertainty of several million years or more; this is one reason – there are several – why a precise concordance between crater ages and extinction times does not exist. Some cosmic objects, as at the end of the Cretaceous Period, must have crashed into the oceans, leaving no visible crater. And, as we would expect, there are random craters that are not part of any 30-million-year cycle.

Since these heady announcements were made in the scientific literature, a number of scientists have gone back to the original data to scrutinize the reliability of the conclusions drawn. How complete is the list of craters? How accurately

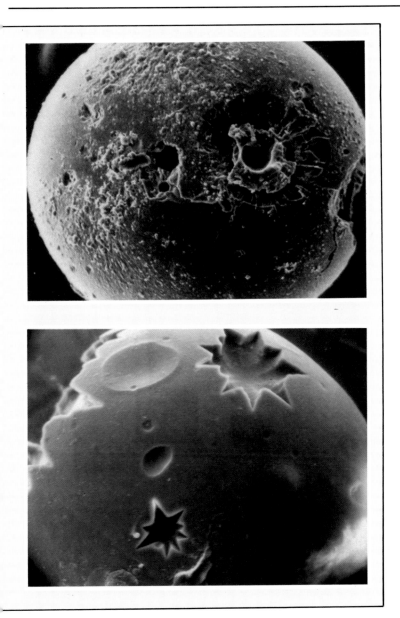

determined are the times of impact? How do you define a mass extinction? How reliable are the dates of cratering or mass extinctions? Suppose cratering and extinction events happened entirely randomly; how likely is it that you would, by chance, find spurious evidence of periodicity? The results of this re-analysis are not yet, in the middle 1980s, fully in, but in the heated debates that have ensued it is clear that the foundations of several sciences are being constructively reassessed. There are those who believe in an in-phase periodicity of cratering and mass extinctions. There are those who believe that cratering may be periodic, but mass extinctions are not, or the reverse. And there are many scientists who hold that, within the present uncertainties,

...The evidence for [periodic mass extinctions] is strongly contingent on arbitrary decisions concerning the absolute dating of stratigraphical boundaries, the culling of the database and the definition of what is mass extinction as opposed to background extinction. This evidence becomes insufficient under other plausible geological timescales and other acceptable definitions of mass extinction. Analysis of the non-culled database shows that the reliability of identification of mass extinctions and their timing is at present extremely limited. It also suggests that the apparent periodicity of mass extinctions results from stochastic [random] processes.

Antoni Hoffman,
'Patterns of Family Extinction Depend on Definition and Geological Timescale' *Nature, 315,* p. 659 (1985).

there is no compelling evidence for periodicities, either in cratering or in mass extinctions.

If this last group is right, then the remainder of this chapter is only a flight of fancy. If there are such periodicities, a stunning connection between life on Earth and cosmic events has been uncovered. In that case, somewhere a cosmic doomsday clock is ticking away even now. Fortunately, we are today midway between mass extinctions. The next bombardment is not scheduled, if the periodicity is correctly determined, for another 15 million years. Our job is to avoid becoming one of the background extinctions; in 15 million years, we may be ready for the comets.

But how could any object in space know when its time had come to hit the Earth? What cosmic machinery could serve as a doomsday clock? Take, for example, the present population of short-period comets, or the hundreds of asteroids that cross the Earth's orbit. They will strike the Earth from time to time, of course, and contribute to the background rate of cratering. Indeed, much of the background iridium in rocks, unassociated with biological catastrophes, may also come from comets – the steady infall from meteor streams and the zodiacal dust. But this will be a steady pitter-patter; these worlds do not save themselves up for mass assaults on the Earth every 30 million years – or, for that matter, on any other timescale. The impacts must be caused by other, more distant bodies.

By the mid-1980s, two quite different proposals were in contention, each purporting to show how a cosmic metronome might beat out extinctions at 30-million-year intervals. Neither is fully satisfying. It is hard even to state the

---

PASSING STARS

Before a periodicity of extinctions and cratering was proposed, J.G. Hills of the Los Alamos National Laboratory suggested that a passing star – rather than a companion star – in the inner Oort Cloud, could stir up a storm of comets:

> The observed comet cloud may be only the outer halo of a much more massive comet cloud whose centre of mass is well inside the observed inner boundary of the Oort Cloud. The possibility of this being the case was already realized by Oort in his classic paper.

Hills then goes on to deduce that a star passing only 3,000 Astronomical Units from the Sun would produce a cometary shower from the inner Oort Cloud that would generate one new comet per hour in the vicinity of the Earth. This would have many consequences, but the first one that Hills, an astronomer, mentions is this:

> Such a high comet flux would be a major nuisance to astronomical observers engaged in research on low-light objects!

This is another of the rare astronomical exclamation marks. Hills then continues:

> The integrated comet flux from such a shower can be great enough that several comets will actually hit the Earth during the shower. This may show up in the geological record.

J.G. Hills,
Astronomical Journal,
Volume 86, p. 1730, (1981).

---

two hypotheses without seeming a little lurid, and the more sensational recountings in the popular press have, in the minds of some scientists, confirmed their initial misgivings. But together, they provide a good example of science in transition. A powerful conclusion is drawn from disputed data. But the conclusion poses an enigma. Two hypotheses attempt to explain the enigma in different ways. If science is to be served, the hypotheses should make different predictions about what you will find if you perform a particular sort of new experiment. Experimental confirmation of quantitative predictions is to science what the fulfilment of prophecy is to religion.

In one of the two hypotheses, life on Earth is periodically extinguished because the Milky Way galaxy has the shape first understood by Wright and Kant (see Chapter 4). Our galaxy is a thin disc that contains the spiral arms, and that rotates about a bulging core of stars and dust (see the frontispiece to this chapter). It is at the centre of the Milky Way that huge numbers of stars are concentrated; it is, in brilliance, mass, position, and explosive violence, the downtown of the galaxy. Fortunately for us, we live nowhere near it, not even in the

16.4 *The motion of the Sun in the galaxy. The Sun orbits the massive, reddish, bright, bulging core of the Milky Way galaxy once every quarter billion years. As the Sun slowly circumnavigates the centre of the galaxy, it bobs up and down at a faster pace, making a bounce roughly every 60 million years and therefore crossing the plane of symmetry of the galaxy roughly once every 30 million years. Does this 30-million-year period somehow trigger mass extinctions on the Earth? (Painting by Jon Lomberg.)*

suburbs. Our home is in an obscure galactic countryside, where stars like the Sun take 250 million years to go once around that distant hub.

But in addition to revolving around the galactic centre, the Sun has another motion: it is bobbing up and down, each time passing through the imaginary plane of symmetry that cuts through the galactic centre. When at its maximum excursion, about 230 light years above the plane, the Sun is gravitationally attracted by the gas and dust and stars beneath, slowly reverses direction, and falls back. But the galactic plane is an imaginary not a real surface, and when the Sun arrives there it finds it has a sizeable velocity and nothing to stop it. Accordingly, it plummets through and out the other side, slowing because of the gravity of the dust and stars it has left behind – until it arrives some 230 light years on the other side of the galactic plane, stops, and then falls back again.

Since the Sun is travelling through a vacuum more nearly perfect than anything we know, there is no friction to retard its motion. Like a perfectly elastic rubber ball dropped on a pavement, the Sun yoyos up and down forever. The oscillating Sun is a perpetual motion machine. As long as a great deal more matter doesn't move into or out of our neighbourhood in space, this galactic bounce will continue until the Sun itself dies. The period of oscillation depends only on how much mass there is in the space near the Sun, and this seems to have been well measured by astronomers. Also, the Sun's motion can be measured by examining the nearby stars with respect to which our system is moving. In both these ways, it is found that the period from one crossing of the galactic plane to the next, for the Sun and its entourage of planets and comets, is about 30 million years, In fact, all the nearby stars are bobbing up and down, crossing the galactic plane once every 30 million years. This period is fundamental for millions of suns. You look at the ages of fossils and craters on Earth, and you find a timescale of roughly 30 million years. You look at the Sun bobbing in the galaxy, and you find another timescale of 30 million years. It's hard not to think that the two periods are related, that the oscillation causes the extinctions.

But how could the Sun's motion instruct comets to collide with the Earth? There are giant molecular clouds spread out in dribs and drabs in our part of the Milky Way. They do not all move at the same speed as the Sun, and, every now and then, the Sun is bound to run into one of them. Extending over a much larger volume than the solar system, they can contain much more mass than the solar system. Passing by or through such a cloud would gravitationally produce a great flurry among the comets of the Oort Cloud, and pry loose a shower of comets to descend on the Earth and its neighbouring worlds.

For this metronome to work, though, the clouds must be concentrated in one particular place – the galactic plane, probably. Then, every 30 million years, the Sun plunges through the plane (alternately from above and from below), a shower of comets is launched towards the inner solar system, one or more of them strikes the Earth, the lights go out and the temperatures drop. If the interstellar clouds lived in the galactic plane, the oscillating Sun explanation of periodic mass extinctions would seem very promising.

But, instead, for the whole 230 light years of the Sun's maximum excursion from the galactic plane, the clouds are distributed essentially at random. There is no more reason to bump into one in the galactic plane than there is 230 light years above or below it. Indeed, at the moment, the solar system is only about 25 light years above (north of) the plane of the galaxy, and there is no giant interstellar cloud at our door. The extinctions should, therefore, occur at random. The 30-million-year period of galactic plane-crossing for the Sun does not seem to translate into a 30-million-year cycle of mass extinctions on Earth. Maybe there's some other way in which the Sun's periodic bobbing in and out of the galactic plane can

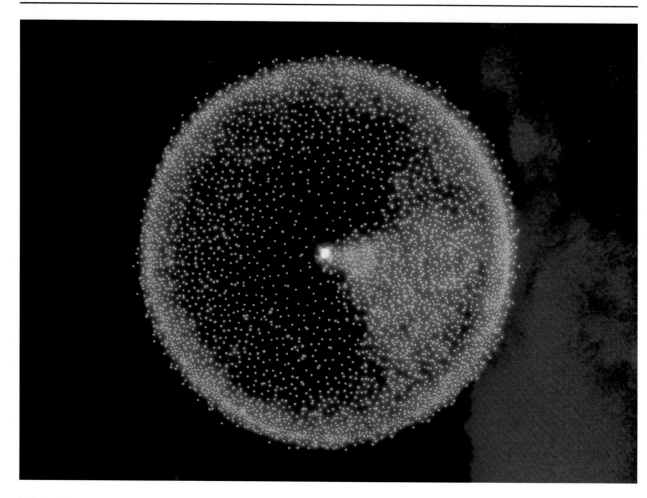

I6.5 *The Sun, at the centre, and its entourage of comets, shown in blue, encounter a massive interstellar cloud. The resulting gravitational perturbation sends a shower of comets into the inner solar system. Such showers must happen statistically from time to time. According to one view, they also happen periodically because of the Sun's motion above and below the galactic plane. (Diagram by Jon Lomberg/BPS.)*

stir the comets up and cause global catastrophes on Earth. Maybe there's a flattened population of small, primordial black holes in the galaxy, through which the Oort Cloud runs as the Sun bobs. But if there is, no one has been able to find it so far; and without something of that sort, this hypothesis, tantalizing as it is, remains unsatisfactory. What is the alternative?

Most of the stars in the sky are members of double or multiple star systems. In a typical binary system, two stars separated by a few Astronomical Units are doing a stately gravitational fandango. Often, the stars are more widely separated. In some instances, we see two stars gravitationally bound to each other, but separated by 10,000 AU. At least fifteen percent of the stars in the sky seem to have a companion star at this distance. The nearest star system to the Sun – Alpha Centauri, 4.3 light years away – is a double star with a distant dim companion, Proxima Centauri, at 10,000 AU from the two bright stars. Often, the companion star is very faint, suggesting that there may be many still-undiscovered, widely-separated binaries. It is possible that

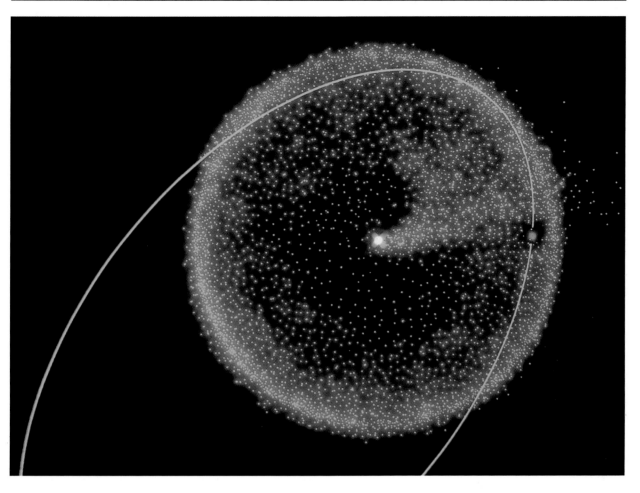

most of the stars in the galaxy are so dim that astronomers call them brown or black dwarfs. Most distant companions might be of this sort.

The solar system seems to be an exception. We do not know of any companion to the Sun. But if we were *not* an exception, if the Sun *had* an invisible companion star in a very specific orbit, then the extinction clockwork again might be understood. Suppose there was a companion star on a very long elliptical orbit – so that, on average, it was 90,000 AU (1.4 light years) away – but once each orbit it comes much closer to the Sun, maybe 20,000 AU, or even a little closer. This would bring the star into the inner part of the Oort Cloud, where comets are not ordinarily jostled by passing stars. With such an orbit, once every 30 million years, the companion would plough through the denser parts of the Oort Cloud, and shower the Earth and neighbouring worlds with comets.

The year of this star would be 30 million of ours. Ten orbits of the Sun would carry it from the Permian, when there were dragons with sails on their backs, to now, when humans reconnoitre the planets and menace the Earth. At present, it

16.6 *The Sun, at the centre, and its entourage of comets, encounters a small, dim star indicated by the red dot. The resulting gravitational perturbation sends a shower of comets into the inner solar system. Such encounters must happen statistically from time to time. In addition, one view holds that the Sun has a companion star on a highly eccentric orbit that enters the Oort Cloud around its perihelion, and therefore periodically sends comets cascading in towards the planets. (Diagram by Jon Lomberg/BPS.)*

would be near aphelion, the furthest point in its orbit from the Sun. Fifteen million years from now, it would run its next rampage among the comets. During its plunge through the inner Oort Cloud, the companion star would spray a billion comets into the inner solar system. But because they would be carried on slightly different trajectories, the comets would not all arrive at the same instant. Rather, they would be spread out over a million years or more. Thus, the companion star hypothesis predicts that the background cratering and extinction rates will be periodically interrupted by intervals lasting as much as millions of years, in which the Earth is hit by several (perhaps several dozen) comets. This would explain why mass extinctions do not occur instantaneously, but sometimes over a period as great as millions of years.

The companion star's period around the Sun cannot be exactly 30 million years; indeed, it cannot be exactly anything: gravitational perturbations by stars and interstellar clouds will tug it first one way and then another, altering its orbital period. After a certain amount of time, any such orbit will be changed greatly. A companion star in the present orbit could survive at most a billion years before being stripped away by passing stars and the general gravitational pull of the Milky Way. This is long enough to account for periodic extinctions and impact craters back to the Permian, if there are any – but, before that, the companion star must have had a very different orbit. If the Sun's companion began much closer to the planetary part of the solar system, it would have produced major and almost continuous cometary showers early in the history of the solar system, and might even be responsible for the 'late heavy bombardment' of the Moon. However, such a greatly enhanced cometary flux early in the history of the solar system is expected anyway, as comets were cleared out from the region around Uranus and Neptune – being propelled both outwards away from the Sun to populate the outer Oort Cloud, and inwards to crater the terrestrial planets (see Chapter 12).

But before we seriously consider such extravagant possibilities, the hypothesis must face one awkward fact: there is, to date, no evidence whatever of a companion star. It need not be very bright or very massive; a star much smaller and dimmer than the Sun would suffice, even a brown or black dwarf – a planet-like body insufficiently massive to generate hydrogen fusion at its core and thus burst forth with stellar light. It is conceivable that the companion already exists in one of the catalogues of dim stars without anyone having noted something anomalous – namely, the enormous apparent motion of the star each year against the background of more distant stars (the effect of parallax – see Chapter 2). There is at least one major search for dark, cool stars now under way and designed specifically to catch a companion star, if it exists. If it *is* found, and in something like the right orbit, few will doubt that it is the principal cause of periodic mass extinctions on Earth. Otherwise, the theory will remain fascinating but unproved.

But it is a notion of mythic power. If an anthropologist of a previous generation had heard such a story from his informants, the resulting scholarly tome would doubtless use words like 'primitive' or 'pre-scientific'. Consider such a story:

> There is another sun in the sky, a demon sun we cannot see. Long ago, even before great grandmother's time, the Demon Sun attacked our Sun. Comets fell, and a terrible winter overtook the Earth. Almost all life was destroyed. The Demon Sun has attacked many times before. It will attack again.

This is why some scientists thought the theory of the death star a joke when they first heard of it. An invisible sun attacking the Earth with comets sounds like delusion or myth. It deserves an additional dollop of scepticism for that reason: we are always in danger of deceiving ourselves. But the theory is serious and respectable, if highly speculative, science, because the principal idea is testable: you find the star and examine its orbital properties. It is no wonder, though, that one of the groups of scientists that first proposed the theory turned to mythology to name the star, and called it Nemesis – after the Greek goddess who visited just punishment on the self-righteous. They then went on to add: 'We worry that if the companion is not found, this paper will be our Nemesis.'

Scientists tend to take these things personally. If there is no Nemesis, some form of the galactic oscillation theory will become more probable, and we will have learned much about dim, nearby stars. Or it will strengthen the case of those who doubt there are periodic extinctions. There is no way that science loses. For the grandeur of its vision, and the new findings that will emerge even if there is no companion star, this theory is in the mainstream of good science, a fact that is sometimes poorly appreciated in the press. The *New York Times,* in an editorial dated 2 April 1985, decided not only to take a stand in this scientific debate, but also decried the fact that discussion of extinctions

> ... was suddenly snatched away by two brash Berkeley scientists and a crowd of astronomers ... Terrestrial events, like volcanic activity or changes in climate or sea level, are the most immediate possible causes of mass extinctions. Astronomers should leave to astrologers the task of seeking the cause of earthly events in the stars.

In a letter dated 14 April 1985, Richard Muller and Walter

Alvarez of the University of California at Berkeley replied to the *Times*, concluding:

> You say, 'Complex events seldom have simple explanations.' The entire history of physics contradicts you.
>
> You suggest, 'Astronomers should leave to astrologers the task of seeking the cause of earthly events in the stars.' May we suggest it might be best if editors left to scientists the task of adjudicating scientific questions?

The debate is likely to continue.

As for the name for the death star, should it be found, the American paleontologist Stephen Jay Gould argues that Nemesis is not quite right:

> Nemesis is the personification of righteous anger. She attacks the vain or the powerful, and she works for definite cause ... She represents everything that our new view of mass extinction is struggling to replace – predictable, deterministic causes afflicting those who deserve it ...
>
> Mass extinctions are not unswervingly destructive in the history of life. They are a source of creation as well ... Mass extinction may be the primary and indispensable seed of major changes and shifts in life's history ... Moreover, mass extinction is probably blind to the exquisite adaptation evolved for previous environments of normal times. It strikes at random or by rules that transcend the plans and purposes of any victim.

Gould proposes instead that the companion star be named after the Hindu deity, Siva, the Destroyer, for whom destruction is a prerequisite of creation, and vice versa:

> Siva does not attack specific targets for cause or for punishment. Instead, his placid face records the absolute tranquillity and serenity of a neutral process, directed toward no one, but responsible for the maintenance and order of our world.

How characteristic of us, Gould is saying, that associating comets with 'the wrath of heaven' – which Laplace denounced as superstition – should find its way into the most recent scientific view of the connection between comets and life on Earth.

16.7 *(Opposite) A light year from the Sun, a still undetected and perhaps wholly mythical dwarf companion to the Sun glowers redly, with a hypothetical planet and comets – of which there are trillions thereabout – all feebly illuminated. (Painting by Anne Norcia.)*

A more disastrous situation than a comet shower would arise if Nemesis were perturbed into an orbit that takes it into the planetary system. In this case, it may strip some planets directly from the Solar System. It would leave the remaining planets in highly eccentric orbits which would produce a fast-moving, dangerous situation... A Nemesis with a semi-major axis of 90,000 AU is dangerous to life on Earth, but a Nemesis at the inner edge of the Oort Cloud would probably have produced a catastrophe of truly cosmogonical proportions.

J.G. Hills,
'Dynamical Constraints on the Mass and Perihelion Distance of Nemesis and the Stability of Its Orbit'
Nature, 311, p. 636, (1984).

These foreign planets, with their tails and their beards, have a terrible menacing countenance. It may be they are sent to affront us.

Bernard de Fontenelle,
The Plurality of Worlds (Paris, 1686).

Chapter 17

THE ENCHANTED REGION

We have now advanced to the extreme boundary of the solid ground of our knowledge of comets. Before us lies the enchanted region of speculation.

William Huggins
On Comets,
Friday Evening Discourse,
The Royal Institution
(London, 20 January 1882).

Everything you see around you, alive or inanimate, has fallen from the sky – or, at least, its constituent atoms have. There was a time, before the Earth was formed, when these atoms were parts of small worlds, busily manufacturing planets as they collided. And there was a time before that, when all the atoms in and around you, every one of them, were in microscopic grains floating in interstellar space. After the Earth was fully formed, more cosmic matter, chiefly from comets, glued itself to the Earth's surface. Everything on Earth is sky stuff – collected, for a while at least, on this lump of metal, rock and water.

In the scientific literature, as in myth and folklore, the favoured implications of comets for life tend to be on the dire side. Only occasionally is there a suggestion that comets may be benign. The earliest such statement seems to have been made by Isaac Newton in the *Principia*. He argues that cometary matter accumulates on Earth:

> The vapours which arise from . . . the tails of the comets may meet at last with, and fall into, the atmospheres of the planets by their gravity, and there be condensed and turned into water and humid spirits; and from thence, by a slow heat, pass gradually into the forms of salts, and sulphurs, and tinctures, and mud, and clay, and sand, and stones, and coral, and other terrestrial substances.

There is a clear implication here that comets are made chiefly of water, although no justification of this assertion is offered. Newton goes on to argue that whatever cometary 'vapours' arrive on Earth just balance those that are being lost, presumably to space, and that this contribution is essential for the continuing fecundity of the Earth. He concludes, surprisingly, in an almost mystic reverie:

17.1 *(Opposite) Waters flow briefly on Mars after a cometary impact. (Painting by Don Davis.)*

> I suspect, moreover, that it is chiefly from the comets that spirit comes, which is indeed the smallest but the most subtle and useful part of our air, and so much required to sustain the life of all things with us.

Comets supply, Newton seems to be saying, something essential for life on Earth. But calling it 'spirit' suggests it is not composed of matter. Demonstrating its presence in comets would then seem a considerable undertaking, which Newton does not attempt.

If the comets were to contain something special for life – something that starts it, say, or something that stops it – there would be ample opportunity for that something to take effect. All over the Earth, all the time, there is a fine rain of cometary debris, particles ultimately derived from the outermost reaches of the solar system. They fall into the atmosphere, are carried out by rain and snow, and are incorporated into plants and animals. We are surrounded by cosmic particles in our everyday life. We eat them and breathe them. And we are made of them. Some of these extraterrestrial atoms are incorporated into sperm and egg cells, and are carried down to future human generations.

In addition to the fluffy, organic-rich particles found in the stratosphere (see Chapter 13), there are other, slightly larger tokens of the comets. For example, if you dredge the ocean bottoms and then, after sieving, examine all fine particles smaller than a grain of sand, you will find tiny, almost glassy spheres. They are present on land also, but because the land is now so polluted by fine particles of human manufacture, the glass spheres are harder to find there. Since their appearance is so similar all over the Earth you might think they were the product of some widely distributed small animal – eggs, perhaps, or droppings. But they are made uniformly of a kind of smooth rock, and they are certainly not biological.

These spherules are the remnants of small, originally irregular objects that entered the Earth's atmosphere at high speed, were melted by friction with the air, and slowly fell through the atmosphere. They are called 'meteor ablation spheres' – droplets that have cooled and re-solidified, teardrops falling on the Earth – a little like the microtektites (see Chapter 16), but much more numerous. If you examine them, you find those chemical elements that are easily volatilized to be under-represented. Carbon is such an element, as is sulphur. They have evidently been boiled off during entry. But other, more refractory elements are present in the usual cosmic proportions. The ablation spheres look like small pieces of cometary dust that have melted, but otherwise survived high-speed entry into the Earth's atmosphere. When interplanetary particles collected at stratospheric altitudes are melted experimentally in the laboratory, they take on the physical and chemical properties of the little spheres found on the ocean bottoms, and, more rarely, in the stratosphere itself. And about ten percent of these spheres collected from oceanic sediments contain grains of the parent material that have, by accident, escaped melting. To clinch the argument, the spherules contain radioactive isotopes, such as Manganese 53, that could plausibly be produced only by cosmic rays in space. Between the droplets and the flakes, we are awash in comet stuff.*

Through the effects of wind and water, temperature and the activities of micro-organisms, all this cosmic debris becomes eroded – abraded, broken down, systematically converted into finer and finer grains – and ultimately reduced to individual molecules and atoms. The ponderous, multi-million-year circulation of the Earth's crust and upper mantle carries the remains of cosmic particles deep into the Earth; hundreds of kilometres below our feet, material that once resided in the comets is part of some great convection cell which, tens of millions of years into the future, will surface and create some new continent or mountain range. The comets may seem remote to the point of myth or fable; but, in fact, they have a most lucid reality – they are a part of our planet and ourselves.

Of all the conceivable materials that could have fallen from comets, the most enduring – both in folklore and in science – has been, oddly, germs, the causative agents of disease. The worldwide association of comets with pestilence is striking, transcending cultural differences, and it is tempting to consider whether comets might, in fact and not just in fancy, be the carriers of epidemics. Through most of human history, of course, disease has been rife, so that, purely by chance, the apparition of a comet must occasionally coincide with an outbreak of pestilence somewhere. For example, the belief that the Great Plague of London, as well as the Great Fire, were caused by comets was sufficiently widespread to be discussed by Daniel Defoe in his *A Journal of the Plague Year*. In 1829, a book by T. Forster was published, called *Atmospherical Causes of Epidemic Diseases*, which argued for the same connection.

The idea is remarkably hardy, and has found a curious modern reincarnation. Fred Hoyle is a leading British astrophysicist who has made important contributions to

---

*The ablation spheres are also found in dust samples collected at stratospheric altitudes, but they are dominated by the smaller, more fragile particles which were able to cool rapidly on entry and avoid melting. Fewer of the spheres fall on the surface; but they are comparatively hardy. Down on the ground, it's much easier to find ablation spheres than organic flakes.

INTERNATIONAL COMETARY EXPLORER

In 1978, a satellite called International Sun-Earth Explorer 3 (ISEE 3), designed to study how the solar wind interacts with the magnetic field of the Earth, was placed in an orbit between the Earth and the Sun. Once its mission was essentially over, ISEE 3 was re-deployed to a quite different task. Robert Farquhar of the Goddard Spaceflight Center designed an ingenious orbital manoeuvre by which the spacecraft will now encounter an extremely interesting object called Giacobini-Zinner, an intermittently active short-period comet which recent work suggests is shaped like a rapidly rotating pancake – with its equatorial radius eight times its polar radius. The comet is also the source of the fluffy particles that make up the Draconid, or Giacobinid, meteor stream. It would be a joy to see close-up pictures of Giacobini-Zinner, but that is not on the cards. ISEE 3 does not have a camera.

To get to Giacobini-Zinner, the spacecraft's thrusters altered its orbit to take it on two leisurely transections of that part of the Earth's magnetic field that points, like a comet's tail, away from the Sun; and then on five consecutive close approaches to the Moon, the last of which, in late 1983, took it to 120 kilometres from the lunar surface. A minor malfunction in the spacecraft's small rocket engine could have crashed it into the Moon. But, instead, the cumulative effect of successive passes by the Moon's gravity flung the spacecraft (like a comet from the Oort Cloud passing close by Jupiter) into a very different trajectory, which will intercept that of Comet Giacobini-Zinner on 11 September 1985.

This is almost six months before the spacecraft flotilla reaches the vicinity of Halley's Comet. So you might think that the exercise is mainly political – like the unsuccessful attempt, in July 1969, by the unmanned Soviet Luna 15 spacecraft to return a sample from the Moon a few hours before the American manned Apollo 11 could do so. And you'd be right. But ISEE 3 – now renamed International Cometary Explorer, with the agreeable acronym ICE – will obtain much more useful information than if it had merely continued to languish in the vicinity of Earth. ICE is a testament to the remarkable abilities to tool around the inner solar system that are now at hand – if you know how to use a small rocket motor, the masses of the Moon and planets, and Newton's laws of motion.

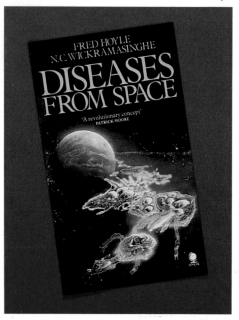

17.2 *The cover of a book by Hoyle and Wickrama-singhe on their postulated cometary and interstellar microbes.*

cosmology, and to our understanding of the synthesis of chemical elements in the stars. He has also written popular science, fiction, and the libretto of an opera. Hoyle and the Sri Lankan astrophysicist N.C. Wickramasinghe have examined the timing and geographical distribution of some cases of epidemic disease, and conclude that the pattern is not that expected if the disease is propagated by contagion on the Earth's surface, the conventional wisdom, but rather is consistent with the pathogens falling from the sky. Most epidemiologists vigorously disagree.

Moreover, Hoyle and Wickramasinghe hold that bacteria and viruses are sprinkled throughout interstellar space;

indeed, they boldly propose that interstellar grains – which have about the same size and atomic composition as bacteria – are, in fact, bacteria. If this is true, bacteria would have been incorporated into comets as they condensed out of the solar nebula, which would itself have been made partly of micro-organisms.

While the sweep of this proposal is very grand, its supporting evidence is sparse. The elaborate molecular machinery employed by both sides in the continuing warfare between infection and immunity speaks of a long evolutionary relationship between humans and disease micro-organisms; this would be hard to arrange if the microbes had recently fallen from a comet that had last seen the inner solar system four and a half billion years ago. The generation of all those interstellar microbes seems difficult to explain, and not all properties of interstellar grains are shared by microbes. There is no report of major epidemics associated – with a one-year time delay – with spectacular meteor showers. In addition, the cometary debris collected by stratospheric aircraft (see Chapter 13) is not reported to contain the charred and mangled bodies of, say, the diphtheria, polio, or cholera pathogens. Scientists examining recovered cometary particles do not seem to be dying of mysterious illnesses. These points must be considered among the deficiencies of the hypothesis.

Furthermore, Hoyle and Wickramasinghe explicitly propose that life has arisen inside the comet (rather than surviving from pre-cometary eons), and is falling on the Earth to generate disease. Here again, the apparent absence of pathogens in recovered cometary grains poses a serious problem. But the idea that conditions inside a comet might be conducive to the origin of life is worth further consideration.

The chief problem is the cold. Comets in the Oort Cloud are so far away that the feeble sunlight is hardly able to warm them at all. They stay far below the freezing point of water or other abundant liquids. But with no liquids (and no atmospheres), it is hard for life to get going: all the molecules are stolid, virtually immobile. The elaborate chemical processing that must have preceded the origin of life on Earth is unlikely to occur at ten degrees above absolute zero. Even if all the required organic building blocks were present, they simply could not interact sufficiently at these temperatures. But sunlight is not the only source of heat.

The interior of the Earth is warm, in part because of a smattering of radioactive elements – uranium, thorium, potassium. Every time one of these elements decays, a gamma ray or a charged particle is ejected, which then strikes neighbouring molecules, mainly silicates, and heats them up. This is, in part, why it is hotter at the bottom of a deep

17.3 *This figure from a German horticultural publication shows an imagined connection between comets and life on Earth: flowering plants and mushrooms are bursting through glass greenhouses to greet Halley's Comet. (Florists' Exchange, 30 April 1910. Courtesy, Ruth S. Freitag, Library of Congress.)*

mine-shaft than it is on the surface of the Earth.

Now, as radioactive atoms decay, they fall to pieces with a statistical regularity. There is a characteristic time, called the half-life, in which half the atoms of a given radioactive species have decayed. The half-lives of uranium, thorium, and potassium are hundreds of millions to billions of years – comparable to the age of the Earth. This is no coincidence: all the radioactive atoms with shorter half-lives have already decayed. So, in the early history of the solar system, there must have been considerably more radioactive heating, and it is more likely that comets were internally warmed as the solar system started out. The bigger the cometary nucleus, the longer it would take for heat to be conducted to the surface and the cooling process to take place. Thus, if early heating is provided by a promising but now extinct radioactive atom called Aluminium 26, it could have kept the core of a comet twenty kilometres across liquid for ten million years; and a two-hundred-kilometre comet for a billion years. Thus, something like the Great Comet of 1729 might have been carrying an ocean inside. A comet with a hundred-kilometre radius, made mostly of water and other liquids, would be enough to cover the entire surface of the Earth to a depth of ten metres. A cometary ocean of such dimensions, rich in organic matter and given four billion years for subsequent chemical interaction, is an interesting sort of place. Is it likely that life could develop there?

The trouble is that everything is happening in the dark. Almost all forms of life on Earth work, one way or another, off sunlight. Ordinary plant photosynthesis is the best-known example. The plants absorb sunlight, and use its energy to convert water and carbon dioxide into carbohydrates and other organic matter. The animals live off the plants. Turn off the lights, and the entire ecosystem collapses. There are a few exceptions to this profound dependence of life on sunlight, but they do not seem relevant to the question of life in a cometary interior. Sunlight is a renewable resource. Not for billions of years will it be all used up. But if life were somehow to arise in such an underground sea, it would gobble up the non-renewable energy sources until they were exhausted. Then there would be nothing left to eat, and the life would become extinct. Just possibly, we will find such a comet at some time in the future, and drill down to the still liquid ocean. It seems unlikely, though, that anything would be alive.

It is much easier to imagine the origin of life on the early Earth, where there was a planet-sized ocean and abundant sunlight. Let us, therefore, leave the unproductive idea that life now or in the beginning fell to Earth from the comets, and ask about

Comets cannot be homes of life; they are not sufficiently condensed; indeed, they are probably but loose congeries of small stones. But even if comets were of planetary size it is clear that life could not be supported on them; water could not remain in the liquid state on a world that rushed from one such extreme of temperature to another.

E. Walter Maunder,
Are the Planets Inhabited?
(1913).

Kazuaki Iwasaki 905 EX

the origin of life from organic molecules on Earth. Life on our planet is built entirely around a handful of molecular types, the most important of which are the nucleic acids and the proteins. If we could understand the large-scale production of these molecules on the early Earth, we would have made significant progress towards understanding the origin of life. The conventional scientific wisdom today – once wildly controversial – is that the key molecular building blocks were spontaneously formed on the early Earth in obedience to well-established laws of physics and chemistry. Molecules in the hydrogen-rich early atmosphere were broken apart by ultraviolet light from the Sun, from lightning, or even from the shock waves produced when meteors entered the atmosphere at high speed. The fragments are known spontaneously to recombine (see diagram) to form the stuff of life. It happens in the laboratory, and it must have happened on the early Earth. Paradoxically, cyanides, the deadly poisons that produced the great comet scare of 1910 (see Chapter 6), seem to be an essential intermediary in the origin of life (see diagram). So could comets have contributed to the origin of life – not by carrying life to the Earth, but by conveying the building blocks out of which life arose?

17.4 *The early Earth at the time of the origin of life. The Moon, much closer then, hangs hugely over the horizon. Organic molecules are carried to the primitive oceans by the infall of cometary matter and through local production in the hydrogen-rich atmosphere by lightning and ultraviolet light. (Painting by Kazuaki Iwasaki.)*

We have seen that in the early history of the Earth – before the inner solar system was swept clean by collision and ejection – comets were much more common; an unmistakable record of these times is preserved in the impact scars on the moon and planets. In the first few hundred million years after the formation of the Earth, the flux of comets was much greater than in the more sedate present. It appears that there were at least thousands of times more comets then, and perhaps still more. These comets, made mainly of frozen water, would, on impacting the primitive Earth, of course have carried water with them. Over the first few hundred million years of Earth history, they would have deposited some 3×10^{15} (3,000,000,000,000,000) tons of water. But this is three percent of the water now in the oceans of the world, enough to cover the whole surface to a depth of almost a hundred metres. If the number of comets in the inner solar system was larger, then the Earth would have accumulated still more water. Just possibly, most of the water in the oceans arrived via a cometary delivery service after the Earth was fully formed.

Certainly, comets would not have been the only source of water. Molten lava arriving at the Earth's surface from the interior is known to carry with it a few percent of water, and volcanic events were frequent billions of years ago. Even so, it is remarkable that comets could have been so important in carrying water to the early Earth.

Not only water would have been brought, because comets contain several percent of complex organic matter. So could cometary organic matter have led to the origin of life? Everything depends upon the timing. If the early comet flux fell off while the Earth was still molten, then no matter how complex the organic matter in the comets, it would have all been fried in the magma; life would have had to originate later, without comets making a significant contribution.

But if the Earth's surface had cooled before the initial pulse of cometary impacts ended, then everything is different. Then, huge cometary nuclei would have fallen on the Earth, their contents of organic matter virtually undisturbed during entry. The comets would have struck the lower atmosphere and surface of the Earth, exploded, and distributed fragments worldwide. Gradually, a warm, diluted soup of organic molecules would have accumulated, both the water and the stock for the broth supplied courtesy of the comets. This is, perhaps, not so different from the view of the Jukun people of Nigeria, who hold that meteors represent a gift of food from one star to another.

A solution of a few percent of organic matter in water is an excellent medium for the origin of life. Indeed, just such a

17.5 *(Opposite) The steps leading to the origin of life on the primitive Earth, according to modern laboratory experiments. Some of the gasses presumed to exist in the primitive atmosphere of the Earth are shown (top left). The hydrogen atoms are the small circles with yellow halos. Molecules shown (upper left) are water vapour (H_2O), methane (CH_4), and ammonia (NH_3). These molecules are broken into pieces (top right) by ultraviolet light from the Sun or electrical discharges. The fragments recombine to form, among other molecules, hydrogen cyanide (HCN) and formaldehyde (OCH_2). The re-combination of these molecules in the ammoniated waters of the primitive Earth produces the simplest amino acids (lower right), the building blocks of protein. Similar steps lead to the building blocks of the nucleic acids. Molecules much more complex than those shown in this diagram have been generated in experiments simulating the origin of life on the early Earth. (Diagram by Jon Lomberg/BPS.)*

dilution is imagined from entirely indigenous processes – the release of water from our planet's interior, forming the oceans; the irradiation of hydrogen-rich gasses in the atmosphere; the slow build-up of organic molecules in the waters of the primitive Earth; and their subsequent interaction leading to the origin of life. Thus we have two competing processes – one from the inside, the other from the outside – both of which seem to lead, at roughly comparable rates, to the complex organic chemistry necessary for the origin of life.

You make a planet like the Earth and it will heat up all by itself and release gasses. The water will make ponds and pools and lakes. If the rest of the atmosphere contains hydrogen-rich gasses and the Sun is shining, huge quantities of organic matter will eventually be mixed with the water. If a comet never fell on the Earth in all the history of the planet, it would still be easy to understand where the original organic matter came from. And if the Earth never released gasses at all, comets would bring an atmosphere, an ocean and huge quantities of organic matter. Thus, in seeking the source of the organic molecules from which we come, we are in the embarrassing position of having two different, and apparently equally successful, hypotheses. It is a challenging task to devise an experiment that can distinguish between them.

If the Earth's primitive atmosphere contained significant quantities of hydrogen-rich gasses – CH_4, NH_3, H_2O, and the like – then an indigenous origin of life can fairly readily be understood. But if (and there is a school of thought that holds to this view) the early atmosphere was not hydrogen-rich – containing, say, N_2, CO_2, and H_2O – then it is difficult to understand the synthesis of organic molecules needed on the primitive Earth for the origin of life, and comets suddenly offer an attractive alternative. Even if indigenous syntheses were entirely adequate, a significant contribution would nevertheless have been provided by the comets. The recipe for life would be: take a million comets, warm gently, and irradiate.

There is nothing in this argument unique to the Earth. Other nearby worlds, in *their* early histories, must also have been pummelled by the comets. Oceans must have accumulated on Venus and Mars, too, even in the unlikely circumstance that no water from their interiors reached the surfaces of these planets. Indeed, for both Mars and Venus, there is some evidence of ancient oceans.

On Venus, a world now depleted of water, the little hydrogen still left is swiftly escaping into space. Lightweight molecules escape more readily than heavier ones, because they are more likely to be induced, by random collision, to move in excess of the escape velocity. The heavy form of hydrogen, deuterium, escapes more slowly than the more

17.6 *Layered terrain at the edge of the Martian polar cap. The Martian polar snows are made of both frozen water and frozen carbon dioxide. (Viking Orbiter photograph. Courtesy, NASA.)*

common, lighter variety. As time goes on, the amount of deuterium relative to hydrogen on Venus should increase. Thus, from the present proportions of hydrogen to deuterium, it is possible to estimate how much water was once present. In this way, Thomas Donahue, of the University of Michigan, and his colleagues have deduced an ancient ocean on Venus, now lost.

Today, Venus is a desolate world, its surface temperature raised to a broiling 480°C by a huge atmospheric greenhouse effect dominated by carbon dioxide. But this CO_2 was not released into the atmosphere overnight, and it is possible to imagine a much more clement early environment, with oceans covering the surface, and dissolved cometary organic molecules colliding, interacting, growing still more complex. The end of the high early flux of comets and the release of carbon dioxide from the interior may have changed Venus from a tropical paradise into an inferno. An important question for future exploration remains: is there any trace today of the ancient oceans? Is it possible that life arose on Venus billions of years ago, with some durable remnant or fossil awaiting future explorers?

These questions can be asked with still greater seriousness for Mars, because, there, a range of evidence exists for abundant water present today – frozen in the polar caps, buried beneath the surface, or chemically combined with the soil. Although liquid water is absent from Mars now, the data suggest that water ran in rivers and floodplains a billion years ago (see Figure 17.7). It is even possible that evidence of shorelines and other signatures of ancient Martian oceans exists in images from the Viking orbiters. Mars is a smaller world than Venus or Earth, so significant amounts of its early atmosphere could have escaped into space. It is also a colder world, so much of the surviving cometary water may now be locked away as ice. Again, the question arises: if there once were extensive oceans on Mars, possibly of cometary origin, did life arise there? Perhaps traces will be found by some twenty-first century expedition.

The worlds in this part of the solar system show the tokens and remnants of the great flux of comets that filled the inner solar system in those long-gone times. We can think of the present modest infall of cometary debris as a feeble remnant of an epoch which lasted for hundreds of millions of years, and which changed the face of the solar system.

If something close to the present oceans were carried by comets to the primitive Earth, simple calculations show that they would also have brought an amount of carbon comparable to that in the entire sedimentary column of the Earth – all the carbon in the rocks for kilometres down, all the

17.7 *An ancient river valley snakes across the battered and cratered Martian landscape. (Viking Orbiter photograph. Courtesy, NASA.)*

17.8 *Layered terrain on Mars, showing signs of a complex and episodic (perhaps periodic) geology. One day, there will be machines and people examining these sediments to better understand the past history of Mars. (Viking Orbiter photograph. Courtesy, NASA.)*

17.9 *A Viking 2 Lander photograph of the terrain in a region of Mars called Utopia. One of the footpads of the spacecraft that landed in 1976 can be seen (bottom right). The metal cylinder above it covered the spacecraft's sampling arm (not in picture) until landing. Note the layer of thin frost that covers much of the picture, ultimately derived, perhaps, from cometary impacts. (Courtesy, NASA.)*

living things, all the humus in the soils, all the petroleum, coal, peat, graphite and diamonds, all over Earth. If anything like this is true, then it is fair to say that the surfaces of the Earth and the other terrestrial planets, once they were almost fully formed, were dusted with a layer of cometary stuff a few kilometres thick. This cometary coating is, relatively speaking, thinner than the confectioners' sugar on a doughnut. But it means the world to us: on this planet, at least, the comet dust has come to life.

The early Earth was covered by impact craters, large and small, and still larger excavations called basins. Some of the water carried by the comets was vaporized on impact, and fell out as rain; the rest reached the ground as intact ice, slowly warming, vaporizing and, again, falling out as rain. So comets dig big holes, in effect fill them with water, and salt them with large quantities of complex organic matter – performing these duties mainly in the earliest history of the planets, so that the maximum time is available for life to arise. Again, nothing in this recounting is restricted to the Earth; the same thing may have happened on countless other worlds in the Milky Way galaxy. If we weren't careful, we might be tempted to anthropomorphize, and think how clever it is of the comets on their own recognizance to take so many important steps towards the origin of life – in this solar system and in countless others.

So take a look around you. If that comet flux near the beginning was thousands of times more than it is now, what comes from the Earth, and what from comets? Everything alive would come from comets – all the plants and animals and microbes, all the men and women. All the buildings, railroads, highways, sculpted farmland, submarines, aircraft, space vehicles – these are all made by humans, and for this reason alone are derived from comets. Even the daytime sky comes from the comets, because O_2 and N_2 are made by life. Well, at least the rocks are from the Earth, we might argue, at least the mountains. Yes, but the rocks are oxidized, chemically altered and broken down by water and by life. The mountains are sculpted and eroded by water. When a lump of the underlying planet tries slowly to poke up through the comet stuff, it is ruthlessly worn down. Even in the wasteland of Antarctica, the landscape appears to be, one way or another, comet-derived. It seems possible that the only things we see that exist entirely apart from comets are the Sun and the stars. A layer of comet stuff once powdered this world and, during the subsequent four and a half billion years, the dust has made a little progress. It has developed complex beings, some aspirations, and a first try at intelligence. Now it is turning its attention to the comets from which it came.

(Overleaf) A comet over Vallis Marineris, Mars. (Painting by Kim Poor.)

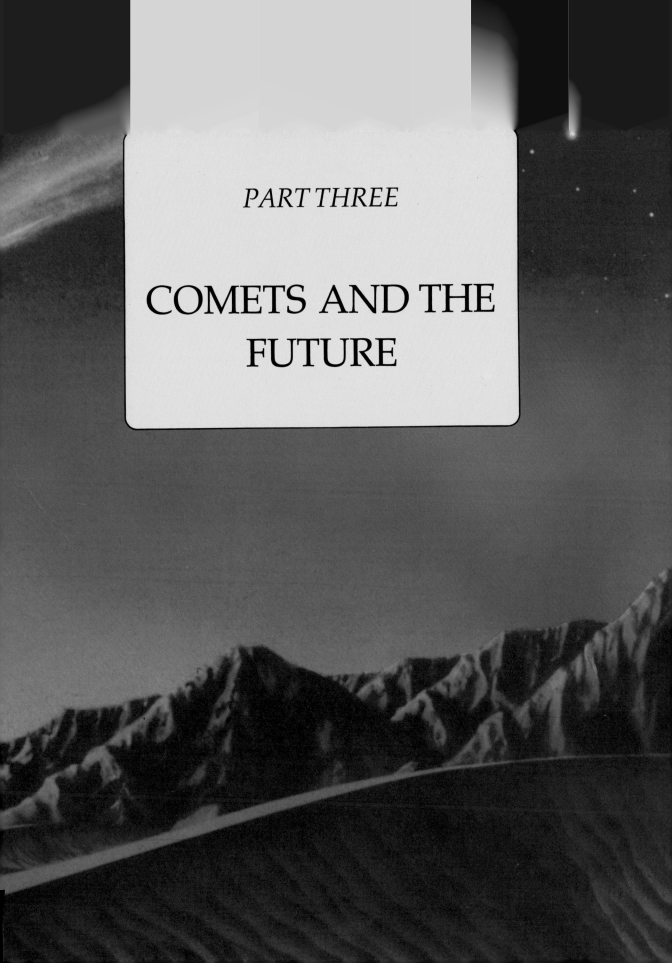

PART THREE

COMETS AND THE FUTURE

Chapter 18

A FLOTILLA RISING

By the middle of the twentieth century, it will have rounded its remotest stake... and will once more have begun the long journey sunward, which will not end 'til 1986. Then again will telescope and camera, spectroscope and photometer, be pointed Halley-ward, just as eagerly as today.

David Todd,
Halley's Comet
(American Book Company,
Cincinatti, 1910).

Shortly before dawn, a rifle shot rang out, and a moment later the whizz of the bullet could be heard, narrowly missing the three occupants of the gondola. Despite the fact that it was May, they were heavily bundled up. In the early hours of the morning, it gets very cold a kilometre or two above the Earth. But they were only 500 metres high when they were fired at. It must have been an unusual sight – the great balloon and its suspended gondola illuminated by the rising Sun when it was still dark in the Connecticut countryside below. Perhaps the balloon had startled some farmer whose only recourse was to shoot it down, whatever it was. Or perhaps the shot was only a safe protest against the idle rich and their expensive diversions.

But this was not a pleasure cruise. Aboard was Dr David Todd, Professor of Astronomy and Navigation and Director of the Observatory at Amherst College. Todd's speciality was the astronomical excursion. During his professional career, he led expeditions to the Dutch East Indies, South America, the Barbary Coast, Russia, Japan, and West Africa – in each case to observe an eclipse of the Sun. He once organized an astronomical foray into the Andes, in part to look for signs of life on Mars. Now, in May 1910, Halley's Comet was in the skies, and David Todd had risen to the occasion – not very far, it is true, but high enough to get above some of the distortion caused by the Earth's atmosphere. He had taken a small, two-and-a-half-inch telescope aboard, with magnification factor of thirty, through which he hoped to observe the comet. The night was clear, and the vibrations from the balloon minor. Accompanying Todd was Charles Glidden, the pilot, and Mabel Loomis Todd, the astronomer's extraordinary wife.

18.1 *(Opposite) A Vega spacecraft about 10,000 kilometres from the nucleus of Halley's Comet in early March 1986. Only a few hundred kilometres from the comet is the tiny form of Giotto, the Halley's Comet explorer of the European Space Agency. (Painting by Rick Sternbach.)*

18.2 'An astronomical observatory in the upper air: taking observations of Halley's Comet from a balloon.' Drawing by Henri Lanos, published in the magazine Graphic, 28 May 1910. (Courtesy, Ruth S. Freitag, Library of Congress.)

A recent historian describes her as follows:

Cheerful, talented, sociable, popular enough to arouse jealous gossip, she was capable of sustaining affectionate and amorous ties; effective as a writer, lecturer, and editor, as wife, hostess, and lover, she constructed for herself, more than most, an enviably robust and resilient character.*

It was Mabel Todd, the daughter as well as the wife of an astronomer, who saw the comet first. The view, her husband said, was much better 'than through the big eighteen-inch telescope at Amherst Observatory' – the observatory he had himself designed and built. The head of the comet was near the horizon, and Todd made four sketches of the tail. This seems to have been the first successful astronomical observation from an artificial platform above the Earth's surface.

Fifteen years later, Todd went on to obtain the first photograph of the solar corona from an aeroplane. But after this success, his behaviour became increasingly erratic, and Todd was intermittently confined to mental institutions until his death at the age of eighty-four, in 1939. Mabel Todd had died before him, mourned by their only child, who remembered especially 'her indefatigable energy and her infectious gaiety'.

Today, new balloons are rising. 1985/1986 represents a historic moment in the study of comets. Previously, the comets came to us. Now, for the first time, we are going to them. The occasion is the apparition of Comet Halley between about November 1985 and April 1986.† Apart from its historic significance (see Chapter 4), Halley's Comet is the only vigorously active comet with an orbit well-enough known to permit detailed scientific planning years in advance. Following David Todd's lead, aircraft will fly above most of the Earth's atmosphere, and rockets will briefly enter space to glimpse the visitor. The unmanned Solar Maximum Observatory, repaired by Shuttle astronauts in April 1984, will examine Comet Halley; special instruments will be flown on the Space Shuttle to examine the comet; and the attention of the US Pioneer-Venus Orbiter will be redirected from Venus to the comet as it passes by. Compared to the usual levels of effort, this would already represent an extraordinarily concerted study. But, in fact, all this represents something of a

*Peter Gay, *Education of the Senses* (New York, Oxford University Press, 1984). Gay, a Durfee Professor of History at Yale University, does not mention this balloon ascent; Mabel Loomis Todd interests him chiefly because of the light her candid diaries shed on Victorian sexual behaviour.

†The remainder of this chapter deals with the forthcoming spacecraft observations of comets for the year or so beginning September 1985. In future editions of *Comet*, this chapter will be fully revised and augmented to take account of the new discoveries.

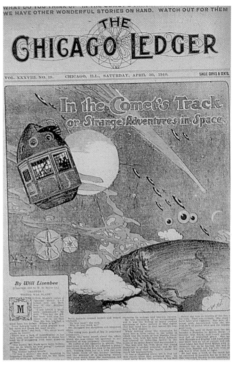

18.3 *Space travel to investigate Halley's Comet as conceived in a Chicago newspaper on 30 April 1910. (Courtesy, Ruth S. Freitag, Library of Congress.)*

18.4 *The Astro 1 mission of the Space Shuttle: specially designed instruments are carried to the Earth's orbit to survey Halley's Comet in early 1986. (Courtesy, NASA.)*

sideshow – because a flotilla of five spacecraft is scheduled to fly by (and one of them, perhaps, into) Halley's Comet in March 1986.

The close-up examination of comets from space is the fulfilment of an astronomer's dream, because we are still painfully ignorant about the most fundamental aspects of comets. We live in the inner solar system, so that cometary nuclei approaching the Earth tend to release gasses and cover themselves with their coma. But if we could fly close to a comet, we might view a cometary nucleus clearly for the first time. What does it look like? What is its shape, appearance, colour? Are there patches of ice, or dark organics, or rock outcroppings? Is it surrounded by a swarm of small boulders? Is there a lag deposit? Are there craters or hills? What about

A comet probe would certainly solve most of the cometary problems without ambiguity.

Pol Swings,
University of Liège,
Belgium, (August 1962).

signs of past surface melting? Is there any hint of layering, as in the Earth's sedimentary column? You might be able to tell a great deal about the nature and evolution of comets if you could photograph a nucleus close up.

Then there is spectroscopy. If we are restricted to the surface of the Earth, we can examine comets only in the visible spectrum, as Huggins did, and through a few 'windows' at infrared and radio frequencies. If we wish to examine comets at other frequencies, we must get above the Earth's absorbing blanket of air. But what can be lifted even into near orbit must be much smaller than major astronomical facilities on the surface of the Earth, so the capability of instruments in the Earth's orbit tends, therefore, to be inferior to that of their ground-based counterparts. Accordingly, only if a comet obliges by coming close to the Earth can instrumentation in the Earth's orbit make important new findings. The discovery of the hydrogen coma around comets (see Chapter 7) is an example; but there have been very few bright comets coming close to the Earth since the advent of space exploration. If you can perform spectroscopy close to, you might be able to discover how different chemical substances are disturbed through the comet.

Generally, spectroscopy performed from Earth reveals the presence of molecular fragments, not the parent molecules in the cometary nucleus from which they come. For example, the nature of the organic molecule from which C_3 derives is entirely unknown. There is an elaborate machinery – involving the release of gasses, the dissociation by ultraviolet light, and further chemistry – that must run its course before we can detect the daughter fragment. Many of these chemical mysteries could be resolved if we could fly close to the nucleus, into the cloud of parent molecules before they are dissociated by sunlight, and there measure directly the molecules, organic and inorganic, that have just been evaporated off the comet.

To detect some sign of the material in the tail of Halley's Comet in 1910, many studies were made at ground level. In France, a large volume of air was brought to very low temperatures, the oxygen and nitrogen was liquefied, and the residuum examined for exotic constituents. Nothing was found. Metal plates coated with glycerines were attached to the struts of a tower at the Mount Wilson Observatory in California, but no cometary dust motes were reported. The experiment is the forerunner of modern studies in which similar glycerine-coated plates are attached to aircraft and flown to stratospheric altitudes – where cometary debris is much easier to come by (see Chapter 13). But now it is possible to fly spacecraft with mass spectrometers aboard that can measure the parent molecules of the comet at first hand.

We think the magnetic field in the solar wind drapes itself over the cometary nucleus, while solar flares produce gusts in the solar wind that generate elegant patterns in the ion tails. But our understanding of comets would be far superior if we flew instruments very near the comet, and measured charged particles and magnetic fields directly.

Visual observers, straining at the limit of the resolution of their telescopes, have detected great fountains of gas and dust shooting into space from the cometary nucleus. We think much of the dust released by comets to space arises in this way. We badly need to sidle up to a comet, photograph the jets – obtaining a motion picture if possible – and fly through the dust cloud, counting and measuring the fine particles.

All this and more is now underway. At the Japanese space centre at Uchinoura in Kogoshima Prefecture, Kyushu, two new launch vehicles of the Mu class lift off. The first is called Sakigake (Japanese for pioneer); the second bears the simple designation, Planet A. They are the first interplanetary vehicles ever launched by Japan.

On the equatorial island of Kourou off French Guiana in South America, an Ariane rocket is launched. In its nose-cone is the first interplanetary spacecraft of the European Space Agency, a consortium of Belgium, Denmark, France, the Federal Republic of Germany, Ireland, Italy, the Netherlands, Spain, Sweden, Switzerland, and the United Kingdom. It is named Giotto – after the Florentine painter who incorporated his own observation of Halley's Comet, during its apparition in the year 1301, into his *Adoration*, a celebrated fresco on the Arena Chapel in Padua.

In Tyuratam in the Kazakh Soviet Socialist Republic, two Proton launchers rise to the heavens. Their ambitious mission is: to fly to Venus – there depositing two spacecraft, to perform the first night landing there, and dropping off two balloon stations to examine the meteorology of the middle atmosphere* – and then continue on to encounter Halley's Comet eight months later. The spacecraft are called Vega – *Ve* from *Venera*, the Soviet word for Venus, and *ga* from *Galley*, the Russian language having an aspirated g, but no h. There are twelve separate scientific instruments aboard Vega's cometary payload. In addition to Soviet instrumentation, the Vega missions carry equipment from Austria, Bulgaria, Czechoslovakia, the German Democratic Republic, the German Federal Republic, Hungary, Poland... and the United States.

The United States of America has played a key role in spacecraft exploration of the solar system – examining

18.5 *The Planet A spacecraft on a test stand in Japan before launch. (Courtesy, Japanese Space Agency and Astronomical Society of the Pacific.)*

18.6 *Halley's Comet as seen in 1301, depicted as the Star of Bethlehem, in Giotto di Bondone's 'The Adoration of the Magi', a fresco completed in 1304.*

*Tasks successfully carried out in June 1985.

18.7 *A Vega spacecraft on its way to Venus and Halley's Comet. The large sphere at the top contains the Venus atmosphere descent package, including the balloon stations. The blue panels are solar cells for the conversion of sunlight into electricity. (Courtesy, Institute for Cosmic Research, Soviet Academy of Sciences.)*

close-up and for the first time every planet from Mercury to Uranus, and dozens of moons. But the United States does not have a spacecraft setting off for Halley's Comet. Several innovative mission designs were proposed by American scientists and engineers; they would have procured a range of essential data that Giotto, Vega, and the Japanese missions will not. But the proposals were turned down by Democratic and Republican administrations alike. There was not enough money. The United States had more important things to do.

18.8 *The logo of the Vega missions. The national flags of participating nations are shown. (Courtesy, Institute for Cosmic Research, Soviet Academy of Sciences.)*

The cost of a major mission to Comet Halley is about that of a B-1 bomber, and less than that of an MX missile. Perhaps in the year 2061, when Halley's Comet next returns, there will be a different American response. But while the United States has, in this respect, opted out of cometary exploration, the flotilla of five spacecraft from twenty nations represents a stirring response by the human species to this emissary from the depths of space and from the early moments of the solar system.

18.9 *Comet Giacobini-Zinner. (Courtesy, International Halley Watch.)*

Through a bit of orbital *legerdemain*, the United States will nevertheless be the first to examine a comet close-up from a spacecraft. It is a rudimentary craft, obtaining no pictures, no spectra, no compositional information, no dust monitoring, but only data on charged particles and magnetic fields.

There is also an American experiment aboard Vega – there entirely through the individual initiative of John Simpson, a professor of physics at the University of Chicago and a veteran of dozens of US unmanned space missions. At a time when the United States had allowed its agreement on co-operation in space science with the Soviet Union to lapse, out of displeasure with Soviet foreign policy, Simpson designed a novel analyzer of cometary dust. He discussed his device at an ESA meeting in the Netherlands, hoping to have it included in Giotto; less than a month later, however, Simpson was

18.10 *The International Cometary Explorer approaches Comet Giacobini-Zinner in September 1985. (Painting courtesy, NASA.)*

informed by Roald Sagdeev, Director of the Institute for Cosmic Research of the Soviet Academy of Sciences, that the instrument had been accepted for the Vega mission. Simpson had never proposed his instrument for the Vega spacecraft. After obtaining permission from the appropriate American authorities, Simpson put together a new device which employed technology at least a decade old; he did not wish to violate US strictures on 'technology transfer'. When the time came to integrate payloads, Soviet engineers asked Simpson why his instrument did not have a computer microprocessor as all of theirs did. Simpson smiled.

All spacefaring organizations involved in the Halley encounter have pledged to make the results available to scientists worldwide, a tradition which has always contributed to the success of science.

The missions have been jointly organized – not only so that the various scientific instruments complement each other and the data can be exchanged rapidly, but also, and much more interestingly, so that the results of one mission will help to assure the success of another.

The most modest are the Japanese missions. Sakigake is considered only a test of the machinery for getting there, and

FORESTED COMETS

I propose to you then an optimistic view of the Galaxy as an abode of life. Countless millions of comets are out there, amply supplied with water, carbon, and nitrogen, the basic constituents of living cells. We see when they fall close to the sun that they contain all the common elements necessary to our existence. They lack only two essential requirements for human settlement, namely warmth and air. And now biological engineering will come to our rescue. We shall learn to grow trees on comets... from a comet of 10-mile diameter, trees can grow out for hundreds of miles, collecting the energy of sunlight from an area thousands of times as large as the area of the comet itself. Seen from far away, the comet will look like a small potato sprouting an immense growth of stems and foliage. When man comes to live on the comets, he will find himself returning to the arboreal existence of his ancestors.

We shall bring to the comets not only trees but a great variety of other flora and fauna to create for ourselves an environment as beautiful as ever existed on Earth. Perhaps we shall teach our plants to make seeds which will sail out across the ocean of space to propagate life upon comets still unvisited by man.

Freeman Dyson,
'The World, The Flesh, and The Devil', Third
J.D. Bernal Lecture, delivered at Birkbeck
College, London, reprinted in C. Sagan, ed.,
*Communication with Extraterrestrial
Intelligence, (CETI)* (MIT Press, Cambridge,
Massachusetts, 1973).

will not pass closer than within a million kilometres of Halley's Comet. Planet A, on the other hand, will approach to within about 200,000 kilometres, and includes an ultraviolet television camera to photograph the hydrogen coma for a month or more before closest approach. Since the hydrogen is produced by dissociated cometary water ice, we will have a record of what gasses are released from the main cometary volatile.

The orbit of Halley's Comet is inclined by 162° to the plane of the zodiac, or ecliptic, that contains the Earth's orbit. Thus, the comet can be intercepted only when its orbit and the orbit of the Earth intersect. For this reason, the trajectories of the Vega spacecraft take them by Halley's Comet on 6 March and 9 March 1986. The closest approach of Vega 1 is designed to be about 10,000 kilometres; it will perhaps be a little less for Vega 2. Giotto is designed to encounter the nucleus of Halley's Comet on 13 March 1986, and pass a few hundred kilometres above it. Because of the enormous relative velocity, the encounter will last only a few hours, and some of the key measurements only a few minutes. But to pass a few hundred kilometres from the comet, you have to know where it is to within a few hundred kilometres, and nobody yet knows its

18.11 *The Planet A spacecraft remains a discreet distance from Halley's Comet while it photographs its hydrogen coma in the ultraviolet. The halo around the Sun is the solar corona. (Painting by Kazuaki Iwasaki.)*

orbit with that degree of accuracy. Accordingly, a co-operative navigation scheme between the Soviet Union, the United States, and the European Space Agency has been organized. It is called 'Pathfinder'. Radio telescopes of the US National Aeronautics and Space Administration will listen, with Soviet co-operation, to the radio transmission from the Vega vehicles, against the background of much more distant extragalactic quasars.*

The position and motion of the spacecraft relative to the Earth can thus be determined with high accuracy. From the first Vega pictures, Soviet space scientists will know to a high degree of accuracy the direction of the nucleus of Halley's Comet. But if we know where Vega is relative to the Earth, and where Halley's Comet is relative to Vega, we know where Halley's Comet is relative to the Earth. This information must be extracted from the observations very rapidly, so that Giotto will be able to effect a most delicate manoeuvre: to optimize the fine detail that the Giotto cameras will see in the cometary nucleus, the spacecraft must pass very close to the comet; but if it passes on the night-side, there will be nothing to see.

*The French/Soviet balloons released into the atmosphere of Venus were also tracked by American radio telescopes. They lasted for about two days before they were destroyed by the fierce environment of Venus.

Without the co-operation of American radio telescopes and Soviet spacecraft, it would be impossible for the Giotto scientists to plan their close flight path, and they might miss by 1,000 kilometres or more. But there is only a couple of days for the Vega pointing data to be received and used to calculate a revised trajectory for Giotto. This will be a race against both time and the comet.

At a 500-kilometre range, the smallest detail visible will be about thirty metres; if pictures are taken at closer ranges, the resolution will be better. Since Giotto will fly as close as possible to the day-side of the nucleus of Halley's Comet, there is then a real possibility that the spacecraft will hit the comet. If the impact occurs, Giotto will be splattered all over the cometary landscape – the relative velocity of spacecraft and comet is about seventy-five kilometres per second (168,000 miles per hour). If there is such a collision, a few scattered fragments of a machine from Earth will be imbedded in the cometary snows and carried out past the orbit of Pluto – like Ahab's corpse, inadvertently marooned on Moby Dick. It would return in another seventy-six years, perhaps to be recovered by future historians of technology.

Even if Giotto does not strike the nucleus, there are other dangers. Ordinarily, the outer skin of a spacecraft is only a few millimetres thick; with no other provisions, such a skin would doubtless be punctured near a cometary nucleus. Radar evidence suggests that comets may be surrounded by a cloud of debris, with some individual particles the size of a fist; and there will be many more small particles than large ones. For safety, the spacecraft has a meteor bumper, of the sort first proposed in the 1950s by Fred Whipple, the father of the dirty ice model of the cometary nucleus. Whipple advocated a thin outer skin surrounding a thicker shield, the two layers in tandem serving to protect the spacecraft from all but the largest particles. The Giotto meteor bumper is made of cloth and epoxy, the same material used for bullet-proof vests. The two Vega spacecraft have something similar.

International co-operation is involved in the Halley encounter in yet another way. Suppose you are interested in, say, an hour-by-hour record of the jetting fountains in a nucleus, or the structure of the tail for months around perihelion. This is simply too much data for the spacecraft to handle, and the observations will have to be done from the Earth's surface. But the Earth turns, the comet rises and sets, and you must have observers distributed all over the planet. More than 900 professional astronomers from forty-seven countries, and many amateurs, have formed an organization called the International Halley Watch, designed to co-ordinate observations. First conceived by Louis Friedman, Executive

Director of The Planetary Society – the largest space interest group on Earth – the International Halley Watch is the first major organized worldwide effort to observe a comet. However, international co-operation has been central to the understanding of comets from the beginning, and Newton's calculation of the trajectory of the comet of 1680 relied fundamentally upon observations made in several nations (see Chapter 3).

Both the Vega and the Giotto television cameras will produce colour photographs. In addition, they have filters that are centred on the frequencies of specific cometary emissions – C_2, for example, and CN (see Chapter 8). We will have pictures of the comet, at close range, in the light of these molecular fragments. William Huggins would have been pleased.

Scientists hope, from photographs obtained far from the encounter, to map the positions of the spouting nuclear jets and position the cameras so they can photograph between the jets. The situation is like seeing the house down the road through a momentary clearing in a snowstorm – but will probably not be as difficult. The nucleus, coma and tail will be examined at many frequencies through the ultraviolet, visible, and infrared. Dust particles will be counted and measured. Mass spectrometers will determine something of the nature of the parent molecules. Charged particles and magnetic fields will be monitored. It seems likely that we are on the verge of a revolution in our understanding of the nature of comets. But if the past history of space exploration is any guide, the most interesting findings will provide answers to questions we have not yet asked.

18.12 *(Opposite) The Giotto spacecraft gingerly approaches the nucleus of Halley's Comet. (Courtesy, Astronomical Society of the Pacific and European Space Agency.)*

There was a time – and very recently – when the idea of learning the composition of the celestial bodies was considered senseless even by prominent scientists and thinkers. That time has now passed. The idea of a closer, direct, study of the universe will today, I believe, appear still wilder. To step out onto the soil of asteroids, to lift with your hand a stone on the moon, to set up moving stations in ethereal space, and establish living rings around the Earth, the moon, the Sun, to observe Mars from a distance of several tens of versts, to land on its satellites and even on the surface of Mars – what could be more extravagant! However, it is only with the advent of reaction vehicles [rockets] that a new and great era in astronomy will begin, the epoch of a careful study of the sky.

Konstantin Tsiolkovsky,
Investigation of World Spaces by Reaction Vehicles
(Moscow, 1911).

Chapter 19

STARS OF THE GREAT CAPTAINS

Whatever opinions we may adopt as to the physical constitution of comets, we must admit that they serve some grand and important purpose in the economy of the universe; for we cannot suppose that the Almighty has created such an immense number of bodies, and set them in rapid motion according to established laws, without an end worthy of his perfections, and, on the whole, beneficial to the inhabitants of the system through which they move.

Thomas Dick,
The Sidereal Heavens and Other Subjects Connected With Astronomy, As Illustrative of the Character of the Deity and of an Infinity of Worlds
(Philadelphia 1850).

When humans discover something new, there is a natural inclination – probably hard-wired into the brain – to put it to use. This odd proclivity, shared in a systematic way by none of the other beasts and vegetables on Earth, is both a principal cause of human success and a major reason that so much of the surface of the Earth is being despoiled. Birds know, better than humans, not to contaminate the nest.

If the trends continue, and if we do not destroy ourselves first, it seems likely that, some time in the twenty-first century, we will begin utilizing the comets. The environment of empty space, the surfaces of large worlds, and the entire volumes of small ones will be visited and made use of by future explorers, both robot and human. The resources, the isolation, and the perspective provided by this new arena may have a benign influence on the quarrelsome human species; but it is at least equally possible that the most irresponsible environmental ethics will be extended to the cosmos. There are bags of human excrement lying on and splattered over the Moon, a vivid symbol of the alternative implications of space technology. Star Wars, the so-called Strategic Defense Initiative, is another.

One of the attractions of the comets is that there are so many of them. Even if the human tendency to litter, contaminate, and obliterate were to continue to increase, it would not

19.1 *(Opposite) Astronauts of the twenty-first century examine the surface of a cometary nucleus. A jet of volatilized ices shoots off into space from beyond the horizon, as the visitors' spacecraft keeps pace with the comet's motion. (Painting by Pam Lee.)*

readily be possible to despoil trillions of small worlds, or even the mere thousands that exist in the planetary part of the solar system. Perhaps we can exercise our ethical muscles more effectively on the comets than on the Earth. However, it would be wholly foolish to begin 'utilizing' the comets – whatever that might entail – without first understanding them thoroughly. After that, an argument can even be made for grinding up and reprocessing whole comets – because in tens of thousands of years, most of the short-period comets will be ejected from the solar system anyway, if they do not first collide with the planets or disintegrate, or transform themselves into asteroids. So there is a certain justification here – if not in the more *outré* subject of battlefield nuclear tactics – for the admonition, 'Use 'em or lose 'em'.

How could you 'use' a comet? You could destroy it or change its orbit if it posed a threat to the Earth; you could exploit it for its resources; or you could colonize it. In this chapter, we discuss each of these possibilities. But one of the earliest practical connections between comets and spaceflight was quite different. Konstantin Tsiolkovsky was a gifted Russian schoolmaster who lived when balloons went higher than rockets. In a series of writings around the turn of the century, he outlined on a grand scale how the rocket engine would eventually carry the human species into space. His influence on the Soviet space programme is still palpable, and a heroic bust of Tsiolkovsky adorns the lobby of the Institute for Cosmic Research of the Soviet Academy of Sciences in Moscow. In a 1911 publication called *Investigation of Cosmic Space by Reaction Vehicles,* Tsiolkovsky wrote:

> Comets have long been expected to bring an end to the Earth, and not without reason, though the probability of this end is extraordinarily small; still and all, it might happen tomorrow or in trillions of years. [But] it will be rather difficult for a comet and other accidental, highly improbable but terrible and unexpected enemies of living beings to strike down in one blow all creatures that have formed [extraterrestrial] habitations...

A comet might conceivably wipe out life on one planet, Tsiolkovsky is saying, but not on many simultaneously; therefore, to ensure the long-term survival of the human species, we should colonize the solar system. The argument loses some of its force, however, if it turns out to be true that cometary catastrophes probably arise from comet showers – whether induced by sporadic passages of stars or interstellar clouds, or by Nemesis, should it exist. Imagine the astronomers on a dozen adjacent worlds tracking some red dwarf star which raises havoc in the inner Oort Cloud, and a swarm of comets – more than enough to impact each of the inhabited worlds – is detected approaching the planets.

However, the general point that Tsiolkovsky makes is certainly valid: the more worlds on which there is a self-sustaining human presence, the less likely it is that random catastrophe will render our species extinct. Tsiolkovsky had not envisaged that humans might, two generations after his time, be able to use the rocket motors about which he was so optimistic to destroy civilization. The history of rocketry has many such ironies. Tsiolkovsky's argument for colonies in space will doubtless be employed to argue that nuclear war need not be so bad as some alarmists say.

The idea that humans should be able to do something about a comet on collision trajectory with the Earth has many antecedents in science fiction, as does the notion of capturing an asteroid or a comet and employing it for human benefit. In *The Comet Hunters,* an early twentieth-century novel by Jean Kerouan, a Professor Grainger invents a raygun to deflect a fictitious Comet Swanley into the Earth's orbit, whence a range of benign climatic and agricultural consequences will ensue. But his treacherous assistant steals the plans – as well as a small bundle of cash – from an American billionaire, adopts the name Kahn Zagan, and assembles the raygun on the Tibetan border of his native Mongolia. One thing leads to another, and Kahn Zagan holds the Earth to ransom, threatening otherwise to deflect the comet so it will impact the Earth. But Grainger and a handy Tibetan Lama, adept in soul transfer, foil the evil plot, and re-route the comet to the Moon – which it hits with some force, rousing our natural satellite from a long geological lethargy. Everything works out satisfactorily – except, of course, for Kahn Zagan. The story is more or less typical of pre-1940 pulp science fiction.

Something similar was described in 1917 in a novel called *The Second Moon* by R.W. Wood and A. Train (Wood was professor of optics at Johns Hopkins University, and a leading physicist of the day). They imagined an asteroid called 'Medusa', whose orbit is by chance perturbed by a near encounter with a comet into a trajectory which would impact the United States. A spacecraft is promptly dispatched, with a newly-invented disintegrating raygun on board, which converts Medusa into a ball of fire, no less. Its trajectory obligingly altered, what remains of the asteroid retreats demurely into a safe orbit about the Earth, where it can be gainfully employed.

Every 10,000 years or so, a comet or asteroid that is around a hundred metres across strikes the Earth. The impact energy is about ten megatons, the equivalent of a single large nuclear weapon, and about a thousand times more powerful than the bomb that destroyed Hiroshima. Because of the absence of

19.2 *A science fiction depiction of a comet in the Saturn system from a 1936 issue of one of the leading science fiction magazines of the time. (Courtesy, Science Fiction Library, Massachusetts Institute of Technology.)*

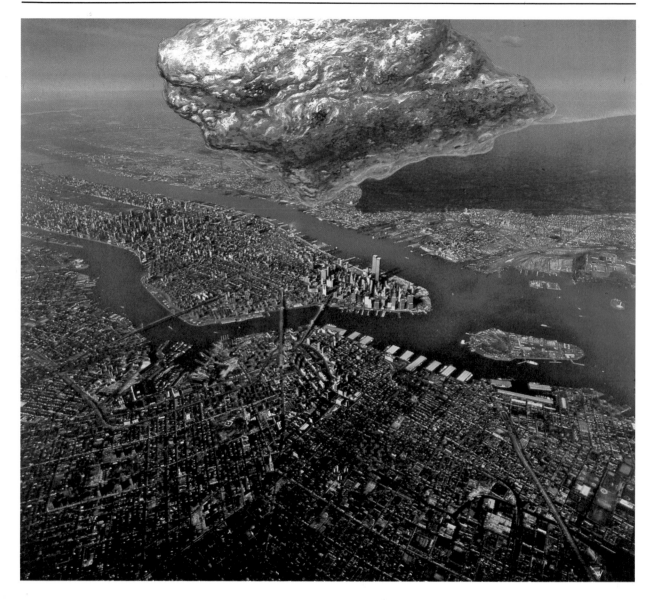

19.3 *A small cometary nucleus over the New York Metropolitan area a moment before impact. (Painting by Michael Carroll.)*

significant radioactivity, the explosion would be much safer than a single groundburst by a nuclear weapon of comparative yield. Impacting randomly on the Earth, it would also be much less likely to fall on a city. The resulting crater, if formed on land, might be as much as a kilometre across. Beyond the crater itself, the damage would be slight. The chance of such a small object falling on a populated area is small, and it is unclear that extraordinary efforts to prepare for such a contingency, or to counter it, are justified.

But once every million years or so, a comet or asteroid a kilometre across will, by chance, run into the Earth. The impact energy would now be about 10,000 megatons – the equivalent of the instantaneous detonation of 10 billion tons of TNT. (If you collected 10 billion tons of a chemical explosive into a cube, the aggregate would be more than a kilometre per side.) This is only a little less than the cumulative yield of all the nuclear weapons that humans seem generally to have sharing the planet with them. (The world arsenals exceed 13,000 megatons.) An explosion of this magnitude would put so many fine particles into the atmosphere that serious climatic consequences might follow, like the impact events that seems to have been responsible (see Chapter 16) for the extinction of the dinosaurs and most species on the Earth 65 million years ago.*

19.4 *A small cometary nucleus improbably makes a direct hit on Manhattan Island. (Painting by Michael Carroll.)*

*Over longer periods, still more catastrophic collisions will occur. Impacts of objects ten kilometres across, the size of a large cometary nucleus, will happen roughly once every hundred million years, producing an explosion a thousand times larger than would result from the simultaneous detonation in a given place of all the nuclear weapons now on Earth.

19.5 *Intrepid but involuntary visitors to a comet from the Jules Verne novel* Hector Servadac.

If such an event happens, on average, once every million years, there is one chance in a million that it will happen next year. But one chance in a million is roughly the risk you take on a commercial airline flight. Many people take such risks seriously. Thus, it makes sense first to refine our census of nearby comets and Earth-approaching asteroids that might one day impact the Earth; and second, to see whether, with existing technology, we might contrive a means to deflect or destroy any object that we discover to have an uncomfortably high probability of impacting the Earth.

Such objects are searched for with ground-based wide-angle telescopes, such as the Schmitt telescopes at Palomar Observatory in California, where the American astronomer Eleanor Helin is now discovering them at the rate of several a year. But such research is woefully underfunded. There are probably a thousand objects around a kilometre in diameter, of which only a few dozen are known about; and a hundred thousand around a hundred metres in diameter, of which only a few are known about – and all of them discovered recently. It is not only good science, but also, as Eugene Shoemaker has emphasized, simple prudence to step up greatly such programmes designed to discover the small worlds in the Earth's neighbourhood – with special emphasis on large, dark bodies. (Most of the big bright ones have already been found.)

As for 'doing' something about an errant comet or asteroid, some provocative studies have been performed. The pioneering work resulted from a student project at the Massachusetts Institute of Technology in 1967, in a graduate course called 'Advanced Space Systems Engineering'. The problem was Icarus, an Earth-approaching asteroid that was scheduled, the following year, to pass within some six million kilometres of the Earth. Suppose, it was posited, that Icarus had instead been making a beeline for the Earth. What could we do about it? Subsequent studies have confirmed the general conclusions of 'Project Icarus'. Atypically for graduate seminars, the joint student/faculty work in this course was published as a book, and led to a high-budget Hollywood motion picture.*

Abandoning the rayguns of an earlier generation, the MIT engineers opted for high-yield nuclear weapons and Saturn 5 rockets. The higher the yield of the explosion, the bigger the resulting crater. But once the depth of the crater is comparable to the size of the intruding object, you've reached a point of diminishing returns; a weapon of still higher yield will only break the object into many fragments. In this way, the MIT

*Called, erroneously, *Meteor*. The word 'meteorite' probably sounded too small-scale.

team estimated that Icarus might be 'destroyed' by an explosion of as few as a hundred megatons, but more likely of a few thousand megatons – a fair fraction of the present global arsenal. However, it is not clear that fragmenting a big comet on a collision trajectory with the Earth into hundreds or thousands of pieces does much for the inhabitants of the target planet. It is a commonplace in science fiction movies for the intruding object to be 'disintegrated' and the threat thereby removed; in reality, all the fragments would continue to move with the original centre of mass, and many would collide with the Earth, spreading the destruction over a much larger area. It makes more sense merely to nudge the comet or asteroid into a new orbit that does not take it dangerously close to the Earth. This also, the MIT group showed, could be done with nuclear weapons.

If you tumble only at the last moment to the reality of an imminent impact, and are required to deflect the intruder on its fateful approaches to Earth, then high yields may be required – perhaps of thousands of megatons: you would emplace your nuclear weapons close together on the surface of the cometary nucleus, and detonate them all at once. The vaporized ice dust and organics would be shot off into space from one side of the nucleus, making a short-lived but effective rocket motor of unprecedented scale. In this way, the comet could be speeded up or slowed down, its orbit changed, and the Earth saved. But it would be better if you had significant advance notice of such a collision. Then, you could finesse the problem during a previous orbit by setting off much smaller explosions at perihelion. A small change in velocity would be multiplied in successive orbits. Doing it this way, the required thrust can be so small that conventional, rather than nuclear, explosives might suffice, or even a real rocket engine – affixed to the offending asteroid or comet and using local material for fuel, oxidizer and reaction mass. This more subtle solution to the problem was discovered after the pioneering MIT study. The lesson seems to be that a little knowledge is worth a great many megatons.

Once we have gone this far, and considered re-routing asteroids and comets to prevent them from impacting the Earth, we can also contemplate rearranging their orbits for other purposes. Before humans invented mining, the only ready source of iron for metallurgy was from meteorites. This fact is still commemorated in many languages, where the word for iron is connected with the word for sky. (The Latin word *siderus* comes from a word for star or constellation, and the Greek word for iron is *sideros*.) Billions of years ago, an asteroid melts, and forms a nickel-iron core under a rocky mantle; a violent collision with another asteroid strips off the

19.6 *Verne's cometary explorers leave the comet as it passes close to the Earth and descend to Europe by balloon. (Frank R. Paul.)*

A comet passed quite near… so they sprang upon it, together with their servants and their instruments.

Voltaire,
Micromegas.

mantle, leaving a bare metallic core, and a subsequent collision sends a shard of iron careening through space. Millions of years later, it falls on the Earth, startling some itinerant tribesperson and providing another inducement toward technology and civilization.

Since the platinum group of metals (see Chapter 17) is relatively more abundant in asteroids (and probably comets) than on the Earth, there may be real economic motivations for mining small bodies that come close to the Earth. An asteroid 100 metres across has a mass of a million tons. There are some 100,000 of them in the vicinity of the Earth. On a planet with dwindling supplies of accessible metals – among them, nickel and the platinum group – this provides significant incentive for commercial utilization of the asteroids. A rocket motor is towed to the asteroid, attached to it, ignited, and used to change its trajectory. Perhaps use will be made of the gravity of the Moon or a nearby planet to bring it to a convenient orbit about the Earth; there, it would be dismantled, and great cargo ships would carry the valuable metals to the Earth.

From a purely economic standpoint, the question is whether the money spent to capture, dismantle, and transport asteroidal material to Earth would be better spent on extracting ores already on Earth, but in less rich deposits or at greater depths. Where the commercial benefit of space mining becomes most apparent is in building large structures in the Earth's orbit. Such constructions might have other justifications: the manufacture of pharmaceuticals and exotic alloys, for example, or for scientific investigations, or, as stations on the way to the planets, and even – if we are so foolish as to permit it to happen – for the introduction of weapons systems into space. It is very expensive to carry material up from the surface of the Earth against the pull of gravity, and far cheaper to utilize materials that are up there already. If we humans begin such large constructions in the next century, it will make enormous sense to use asteroids and comets as raw materials. In the process, we will gain considerable practice in living on small worlds, and moving them about the inner solar system at will. This capability seems likely to open a startling vista.

Crossing the Earth's orbit, as we have said, are roughly a hundred thousand bodies that are about a hundred metres across. Many of them are of cometary origin. The total area of these worlds is ten thousand square kilometres, a sizeable chunk of territory, although, individually, each is only as big as a square block in an urban centre. If there are 100 trillion cometary nuclei in the inner plus outer Oort Clouds, then the total surface area of the comets is the area of hundreds of millions of planets, each the size of the Earth. A hundred million Earths, in equivalent area, parked conveniently in our

Where there are such lands, there should be profitable things without number.

Christopher Columbus,
Journal of the First Voyage
(27 November 1492).

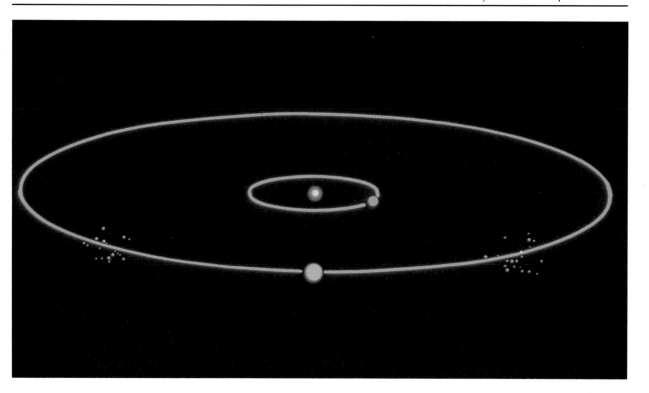

back yard, is a heady prospect.

The comets are travelling so slowly that it would be possible, even with current technology, to overtake them. It takes a comet a million years to get from here to the Oort Cloud; the Voyager spacecraft will do it in ten thousand, and likely future technologies will lower the transit time to less than that of a human lifespan. If we are planning where humans might live in the distant future, the comets offer by far the greatest range of possibilities. But since a few square kilometres will not accommodate very many people, we must imagine a very large number of small worlds, each sparsely populated. But in what sense can a comet be considered habitable?

Certainly, all the molecular necessities of life are to be found on comets. Humans, like most other life forms on Earth, are made largely of water, and except for a few of the moons in the outer solar system, there are no known objects in space richer in water than comets. Comets also contain large quantities of organic molecules useful for agriculture and biological engineering. And we know they probably contain sufficient rock and metal for practical purposes. The great quantities of water also mean that oxygen for breathing should be easily extractable, and rockets working as the Centaur rocket booster does, off hydrogen and liquid oxygen, could be readily refuelled on a cometary surface. In each of these respects, comets are much more obliging bases and habitats than, say, the mainly rocky and metallic asteroids.

19.7 *The Trojan asteroids. Shown here are the orbit of the Earth (background) and the orbit of Jupiter (foreground), both planets circling the Sun. In Jupiter's orbit, 60° ahead of it and 60° behind it are two catchment regions for interplanetary debris, part of a class of relatively stable positions called Lagrangian Points. It is possible that a sizeable collection of both asteroidal and cometary bodies are sitting in these Lagrangian Points awaiting future explorers. (Diagram by Jon Lomberg/BPS.)*

19.8 *You have to bundle up when your comet is as far from the Sun as Jupiter (top left). (From Jules Verne's* Hector Servadac.)

But life on Earth, almost all of it, also runs off the energy of sunlight. The plants harvest the sunlight and the animals harvest the plants. The inner solar system is flooded with sunlight, but, except for the Earth and Mars, is depleted of water. The outer solar system, by contrast, is rich in (frozen) water, but poor in sunlight. Equatorial noon on a cloudless world in the Saturn system is no brighter than twilight on Earth. The water is where the light isn't, and vice versa – a point emphasized many years ago by the American scientist and science writer, Isaac Asimov.

With modern technology, we can now imagine doing something to redress this oversight in solar system design. Comets (and icy boulders from the rings of Saturn) could be driven or towed to the inner solar system, where the ice could be mined directly from the surface, or, for nearby extinct comets, by drilling through the lag deposit to the icy core beneath. The water would be dissociated into hydrogen and oxygen to make rocket fuel and oxidant, and to supply oxygen to human outposts in space and on the terrestrial planets. There is so much water available from comets that it is even possible to imagine water being transported to selected regions of parched worlds, permitting life to be transplanted to previously desolate environments. The organic matter in a dead comet or a carbonaceous asteroid, if finely pulverized, might also be used – as a growth medium for living things, and to moderate the hellish climate of Venus by the same mechanism invoked (see Chapter 16) in studies of the extinction of the dinosaurs and of the Nuclear Winter. Since extinct comets with icy cores are probably already at our doorstep, they may prove to be a critical factor in human utilization of space during the next century or two.

The biological essential that is not supplied by the comet, at least not directly, is heat, warmth, energy, power. These are ordinarily supplied to comets only when they come close to the Sun. We can readily envisage vast arrays of solar panels deployed on and around a comet fairly near the Sun – conceivably even as far away as the orbit of Saturn. Farther out, we can contemplate large nuclear reactors powering cometary bases. If fusion reactors – powered by water itself – become commercially feasible in the middle of the next century, as some experts predict, they would represent an ideal power source for cometary bases, because of the abundance of ordinary water ice, as well as frozen heavy water, HDO and D_2O (where here D stands for deuterium, the form of heavy hydrogen that has a neutron as well as a proton in its nucleus).

A more romantic idea has been proposed by the British-born physicist Freeman Dyson, who suggests that through genetic engineering we will one day be able to design a special tree of unprecedented size to grow on comets far from the Sun. It would be planted in the organic snows and grow enormously, so that its leaves could gather in enough of the sparse sunlight. Various requirements must be fulfilled, including heat insulation, no loss of gasses to the adjacent vacuum, and the like. Because of the low gravity, the growth of the trees is not restricted by their weight, and Dyson foresees forests larger than the comets from which they sprout. The oxygen produced in photosynthesis 'will be transported down to the roots and released into the region where men will live and take their ease among the tree trunks'. 'Will' sounds a little optimistic, but there seems to be nothing impossible about the proposal. But very far from the Sun – within the Oort Cloud, for example – even such heroic measures will be unavailing; the Sun is simply too dim there, and something like a fusion reactor will be needed to power the biological cycles and keep the place tolerably warm.

It is a law of biology, as well as of social relations, that isolation fosters diversity. Picture a time in the far future with millions of inhabited comets, each harbouring no more than a few hundred individuals. Within the Oort Cloud, it would take a day or more for a radio message travelling at the speed of light to reach one colonized comet from another. That might be sufficient to maintain some cultural homogeneity among these many worlds, but the absence of frequent visits would permit a slow divergence of cultural and behavioural norms, and an enormous diversification of social, political, economic, religious, and other views – a development that might be of major benefit for the human species. It is hard to see, though, what advantages might accrue thereby to individual nation states through such activities, and they are currently the only entities wealthy enough to foot the bill. The nation states have a history of preferring their own short-term advantage to the well-being of the species. Thus, the time when the majority of the human species will have been scattered among the comets is, for this and other reasons, far off. But, in the long-term, if space technology continues to develop, we will go to where the surface area, the water, and the organic matter are – to the comets.

If, in the remote future, we populate the comets out to the Oort Cloud, we will have gone, through a succession of slow steps, halfway to the nearest star; and a natural progression exists from there to the rest of the galaxy. The colonization of the galaxy would happen all by itself if the Oort Cloud became populated. Individual comets are so loosely bound that casual gravitational perturbations by passing stars release enormous

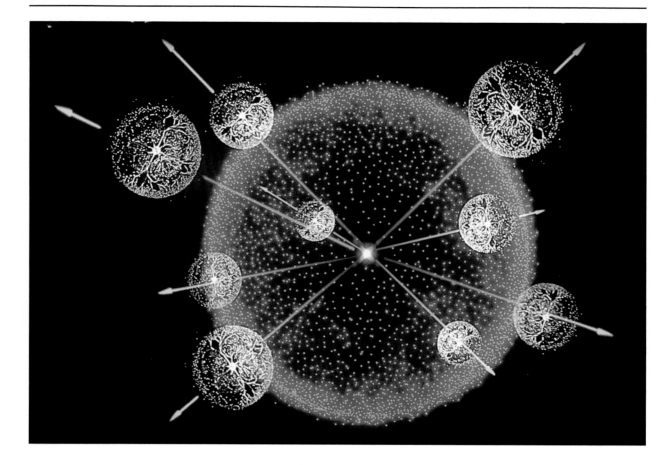

19.9 *The comets of the Oort Cloud have a natural slow leakage to interstellar space, which will perhaps, in the far future, be accelerated by human technology. (Diagram by Jon Lomberg/BPS.)*

What I conceive to be one of the main designs of the Creator in the formation of such a vast number of splendid bodies is, that they may serve as habitations for myriads of intelligent beings... if this position be admitted, then we ought to contemplate the approach of a comet, not as an object of terror or a harbinger of evil, but as a splendid world, of a different construction from ours, conveying millions of happy beings to survey a new region of the Divine empire...

Thomas Dick,
The Sidereal Heavens and Other Subjects Connected With Astonomy, As Illustrative of the Character of the Deity and of an Infinity of Worlds (Philadelphia 1850).

numbers of comets (see Chapters 11 and 16), which then pirouette prettily through interstellar space on their own. In the far future, even if the Sun has no companion star, inhabited comets in the Oort Cloud would be shaken free of the shackles of the Sun's gravity, and would begin seeding at least the nearer parts of the galaxy with humans.

At present, it seems that as much as five Earth masses of cometary material has been ejected from the Oort Cloud since its formation. Much larger masses must have been ejected from the planetary part of the solar system when the vicinity of Jupiter and Saturn was populated by enormous numbers of small, icy worlds (see Chapter 12). Estimates of these masses range from twenty-five to as much as a thousand Earth masses. All this material, originally derived from our neck of the galactic woods, is now tumbling through the space between the stars, randomly perturbed by passing cosmic objects over billions of years of time, and diffusing until they are distributed through the Milky Way.* But up to now, all these comets have been, so far as we know, uninhabited.

*Occasionally, one may briefly enter the inner regions of some other planetary system and streak across alien skies, perhaps provoking some amalgam of fear and wonder in the souls of very different creatures.

No comet has ever been observed on a trajectory originating outside the gravitational influence of the Sun. And yet, sooner or later, such comets should be seen. In our own system, we conclude that many comets have been ejected into interstellar space after close passages by the Sun or the major planets. Especially with the discovery of debris rings around nearby stars (see Chapter 12) it is reasonable to think that many – perhaps most – of the stars in the sky are similarly enveloped in clouds of comets, which are also being ejected into interstellar space.*

Newton imagined comets around other stars:

> This most beautiful system of the sun, planets, and comets, could only proceed from the counsel and dominion of an intelligent and powerful Being. And if the fixed stars are the centres of other like systems, these, being formed by the like wise counsel, must be all subject to the dominion of the One.

And Laplace envisioned comets 'which, moving in hyperbolic orbits, can wander from system to system'. But space is very empty, and the stars are far apart. If every star in the Milky Way galaxy had an Oort Cloud and a cometary ejection rate like ours, the average time between arrivals here of true interstellar comets would be thousands of years. Astronomers are eagerly waiting.

As time goes by, the Milky Way acquires more and more interstellar comets. If every star in the Milky Way ejects a thousand Earth masses of comets into interstellar space every 4.5 billion years, as ours has, then there may be the equivalent of the mass of a hundred million suns floating undetected in the space between the stars. As great a mass as this represents, it is much less than one-tenth of a percent of the mass of the Milky Way itself.

Thus, there seems to be a community of comets filling the galaxy. Comets are likely to form in the accretion disc around every proto-star. If the case of the solar system is typical, every star will eject something like a trillion comets into interstellar space – mainly in the nebular stage, with a trickle of cometary ejections persisting for the lifetime of the star. If there are a few hundred billion stars, each of which has ejected something like a trillion comets, the number of interstellar comets in the galaxy is almost 10^{24} (1,000,000,000,000,000,000,000,000), more than the number of stars in the universe. (The number of comets still bound to stars would be larger.) This multitude of interstellar comets should by now be distributed entirely at

19.10 *Thomas Wright's vision of innumerable suns surrounded by rosettes of cometary orbits. (From his* An Original Theory of the Universe, *1750.)*

*Martin Harwit and E.E. Salpeter, astrophysicists at Cornell University, have proposed that gamma ray bursts, observed by satellites, are due to the impact of comets with neutron stars. If this is true, comets must surround many stars, because these still unexplained bursts come from all over the sky.

random, in and between the galactic spiral arms. The average distance between them, even far from stars, would be some tens of Astronomical Units, the same sort of distance that separates the comets in the Oort Cloud.

The galaxy can then be pictured as a vast, flattened disc of comets, in which are embedded interstellar clouds, stars and their planetary companions. In the vicinity of the stars, the concentration of comets is greater. Accordingly, nowhere in the galaxy would there be a place more distant from a comet than the Earth is from the planets we have already visited. Advanced civilizations may, for all we know, be able to cross the galaxy in one fell swoop on some kind of interstellar express; but the comets provide an opportunity for backward civilizations like ours to build a local train – one that chugs off for a few years, stops at a comet, explores, refurbishes, and then takes off for the next comet. The principal problem is to find and catalogue the nearby interstellar comets. The explorers may have to take a large radar telescope with them.

The comets in the Oort Cloud of the Sun and the cometary clouds of other stars may merge and intermingle, as had long ago been almost anticipated by Thomas Wright (see Figure 19.10). There may even be comets on figure-of-eight orbits, bound alternately to two different stars (see Figure 19.12), if the stars themselves are close enough together. The comets, then, are moving stepping stones to anywhere – like the ice floes on which Liza was able to cross the raging river in *Uncle Tom's Cabin*.

Even if there is no colonization of the comets, we will one day set out to explore the trans-Plutonian spaces, and it will make sense to refuel and refurbish on the comets. They would still be stepping stones to the stars. It is also possible that comets will be themselves converted into spaceships bound for other star systems, taking thousands of generations or more before they approach another sun and the sleeping cometary forest awakens to long-forgotten sunlight. The prospect recalls the image raised by the German astronomer J.H. Lambert, who wrote in the eighteenth century:

Thus we can conceive comets which, being attached to no particular system, are in common to all, and which, roaming from one world to another, make the tour of the universe... I love to figure to myself those travelling globes, peopled with astronomers, who are stationed there for the express purpose of contemplating nature on a large, as we contemplate it on a small scale... we may suppose that their year is measured by the length of their route from one sun to another. Winter falls in the middle of their journey; each passage of a perihelion is the return of summer; each introduction to a new world is the revival of spring; and the period of quitting it is the beginning of their autumn.

'We must not let our people,' said he, 'lose all interest in Earthly matters. They are still the children of the Earth. They are someday to return to the Earth, but even if they never return, it will be useful to attach them to our dear old mother world by at least memory and grateful recollection.'

Jules Verne
Off on a Comet (1878).

19.11 (Opposite) A bifurcated cometary tail is produced in the planetary system of a contact binary, two stars of different properties so close together that matter flows from one to another. (Painting by Don Davis.)

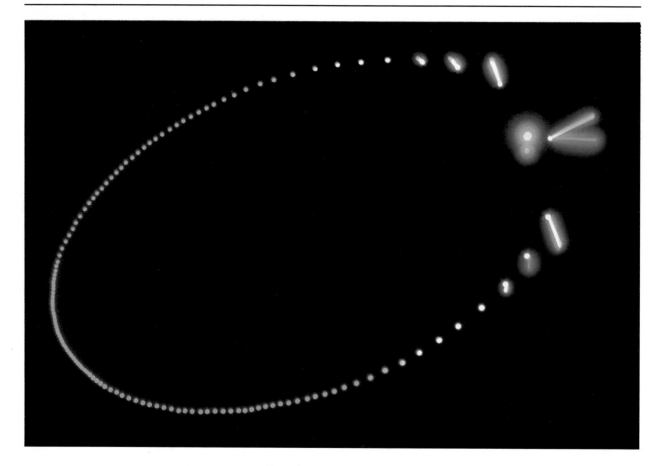

19.12 *A comet orbits a binary star, composed of red and blue component stars. The length, colour, and multiplicity of the tails varies with position in the orbit. (Diagram by Jon Lomberg/BPS.)*

But there is nothing peculiarly human about this story, not even the urge to colonize. We can imagine the comet clouds making every star a dandelion; periodically or episodically, some of the seeds are shaken loose, carrying the local life forms out into the galaxy. Occasionally, a seed travels of its own volition. Sooner or later, it would seem, unless all technological civilizations have an unerring drive for self-destruction, the expanding inhabited cometary clouds will encounter one another somewhere in interstellar space. Perhaps in our explorations of the Oort Cloud, we will find a stray object on which some alien technology is evident.

There are a great many cometary distractions between here and the nearest star. Some of them may be substantial worlds, hundreds or even thousands of kilometres across. If there are 100 trillion comets filling the spaces between adjacent stars, any advanced civilization will expand outwards very slowly. This may be why we have not found evidence of alien visitors in our solar system: the galaxy is too interesting, and they have not discovered us yet.

1985-86 will see, if all goes well, the historic maiden voyages of the human species to the comets. There will be more elaborate

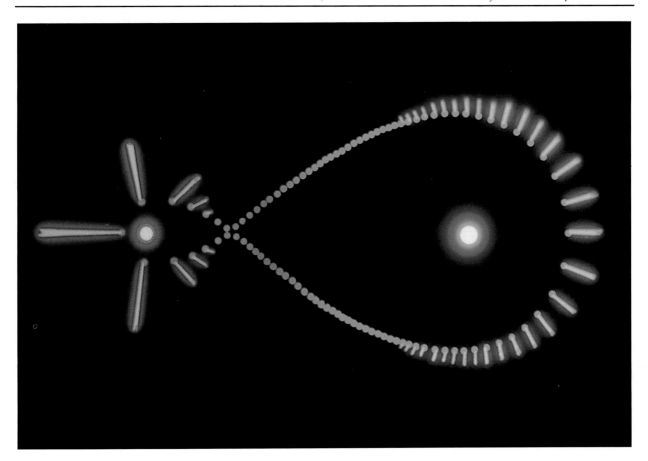

missions to a wider variety of comets, eventually extending to great distances from the Sun. One day – probably in the next century – these spacecraft will carry human crews. We will be living on the comets, and with the aid of rocket engines and Newton's laws of motion, piloting them. If such a day comes, we will have fully justified the faith implied by the !Kung people – that almost unique human culture that considers cometary auspices benign. In the !Kung language, comets are 'The Stars of the Great Captains'.

19.13 *A comet orbits a widely-separated binary star in a figure-of-eight orbit. (Diagram by Jon Lomberg/BPS.)*

Chapter 20

A MOTE OF DUST

Contemplated as one grand whole, astronomy is the most beautiful monument of the human mind; the noblest record of its intelligence. Seduced by the illusions of the senses, and of self-love, man considered himself, for a long time, as the centre of the motion of the celestial bodies, and his pride was justly punished by the vain terrors they inspired. The labour of many ages has at length withdrawn the veil which covered the system. Man appears, upon a small planet, almost imperceptible in the vast extent of the solar system, itself only an insensible point in the immensity of space. The sublime results to which this discovery has led, may console him for the limited space assigned him in the universe.

Pierre Simon, Marquis de Laplace,
System of the World Part I,
Chapter 6 (1799).

A comet is an emissary from the frigid interstellar night that constitutes by far the greatest part of the known universe. And a comet is also a great clock, ticking out decades or geological ages once each perihelion passage, reminding us of the beauty and harmony of the Newtonian universe, and of the daunting insignificance of our place in space and time. If, by chance, the period of a bright comet happens to be the same as a human lifetime, we attribute to it a special and poignant significance:

> Like hundreds of other little boys of the new century, I was held up in my father's arms under the cottonwoods of a cold and leafless spring to see the hurtling emissary of the void. My father told me something then that is one of my earliest and most cherished memories.
>
> 'If you live to be an old man,' he said carefully, fixing my eyes on the midnight spectacle, 'you will see it again. It will come back in 75 years. Remember,' he whispered in my ear, 'I will be gone, but you will see it. All that time it will be travelling in the dark, but somewhere, far out there,' – he swept a hand toward the blue horizon of the plains – 'it will turn back. It is running glittering through millions of miles.'
>
> I tightened my hold on my father's neck and stared uncomprehendingly at the heavens. Once more he spoke against my ear and for us two alone. 'Remember, all you have to do is be careful and wait. You will be seventy-eight or seventy-nine years old. I think you will live to see it – for me,' he whispered a little sadly with the foreknowledge that was part of his nature.*

*Loren Eiseley, *The Invisible Pyramid* (New York, 1970).

It would have been a great satisfaction to know that everyone who saw this wonderful object did so with the same feeling of elation and wonder – one would almost say veneration – with which the average astronomer regarded this beautiful and mysterious object stretching its wonderful stream of light across the sky.

E.E. Barnard, the
pre-eminent observational
astronomer of the time, on
the 1910 apparition of
Comet Halley.

20.1 *(Opposite) From an outpost on the planet Mars, two humans view the return of Halley's Comet in July 2061. The Earth is the bright blue point of light in the sky, and its image on the construction indicates the planet of origin of the settlers. (Painting by Don Davis.)*

20.2 *A late Babylonian tablet, with an account in cuneiform writing of the apparition of Halley's Comet in 164 BC. This is a portion of a systematic Babylonian astronomical and astrological textbook, reading in part, 'The comet which previously had appeared in the east in the path of Anu in the area of the Pleiades and Taurus, to the west... and passed along in the path of Ea'. (Courtesy, The British Museum.)*

20.3 *Two recovery photos of Halley's Comet as it was first seen in 1909 (top, courtesy, Soviet Academy of Sciences) and October of 1982 (above, courtesy, International Halley Watch).*

Halley's Comet is unique in our epoch – a bright, periodic, sometimes spectacular comet that sews the generations together, stitching back through history and forward into epochs to come, awakening us from the delusion that we exist separately from our past and our future. You look up and see Halley's Comet – through a pair of binoculars, perhaps – and you are looking at the same comet* that Edmond Halley saw the summer after he and Mary Tooke were first married. This is the comet whose apparitions were carefully noted by Chinese astronomers. It has been described on tablets, silk, bamboo paper, parchment, newsprint and computer discs. This is the comet that cheered the !Kung hunter-gatherers, and frightened nearly everyone else in cultures far-flung over the surface of the Earth and far back in time – and which continues to arouse suspicious fears to this day. We share this comet with many others.

A million years ago or so, as our hominid ancestors were hunting game and figuring out how to build a house, a passing star sent a gravitational ripple through the cloud of comets that envelops the Sun, and a pristine small world of ice and rock and organics went careening in toward the Sun. Ten or twenty thousand years ago, as our ancestors were dealing with the Wisconsin Ice Age, the comet finally arrived in the planetary part of the solar system, made a close encounter with one of the major planets, and began the conversion into its present orbit – that takes it, once every seventy-six years, closer to the Sun than Venus is, and further from the Sun than Pluto. Every seventy-six years or so, for the last ten thousand years, this comet has swept by, dazzling the inhabitants of the Earth with a cosmic fireworks display, and then vanishing into the night.

Seventy-six years is a few generations or less, and so Halley's Comet is a kind of metronome beating out the rhythm of human progress or decline. Its 1910 apparition was the first since the invention of the aeroplane, and the last before the invention of nuclear weapons. We have lately contrived the means of our self-destruction, and there is a real question as to how many humans will be left next time Halley's Comet comes by the Earth, in July 2061.

The dangers that we face are part and parcel of the process, now well under way, of the unification of the planet – in language, culture, science, and commerce. Sadly, but inevitably – they are both driven by the identical technological advances – this critical and delicate time coincides with the widespread availability of nuclear weapons. At the present rate of change, it seems likely that in the period between now

*Or very nearly the same comet: the nucleus has lost a few metres of ice since the perihelion passage of 1682.

20.4 *Halley's Comet in 1910. (Courtesy, International Halley Watch.)*

20.5 *Six forms taken by Halley's Comet during the 1835 apparition as drawn by John Herschel. (From Amédée Guillemin,* The Heavens, *Paris, 1868.)*

and 2061, the turning point for the human species will have been reached.

If we survive until then, the passage to the following apparition of Halley's Comet should be comparatively easy. That perihelion passage will be in March 2134, in which year the comet will make an unusually close encounter with the Earth. It will come within 0.09 Astronomical Units, or 14 million kilometres, more than twice as close as it did in 1910: it will be brighter than the brightest star. If there are those to do the commemorating, the years 2061 and 2134 should be celebrated for the courage, intelligence, and common purpose of a species forced by urgent necessity to come to its senses.

There are, as everyone knows, contending inclinations within each of us – toward understanding and growth and

It is more probable that [Halley's Comet] has been diverted into its present path, from one of much longer period, at some much less distant time, and has not yet worn out. Its orbit does not now come near any of the great planets, so that its capture can be no recent event; but the accumulation of ordinary small perturbations during a few thousand revolutions would probably suffice to shift the orbit enough to remove all traces of the original point of capture.

Henry Norris Russell,
The Solar System and Its Origin (New York, Macmillan, 1935).

20.6 *(Overleaf) The Comet as a global cultural event. (Courtesy, Ruth S. Freitag, Library of Congress.)*

HALLEY'S COMET SEEN FROM PARIS
Evelyn Paul
178

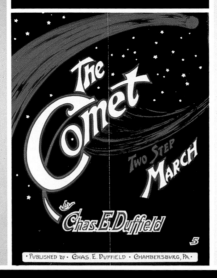

The Comet
TWO STEP MARCH
By Chas. E. Duffield

• PUBLISHED BY • CHAS. E. DUFFIELD • CHAMBERSBURG, PA. •
5

L'ILLUSTRAZIONE
ITALIANA

LA COMETA DI HALLEY.

哈雷彗星今昔

张钰哲著

知识出版社

EL COMETA
HALLEY

COMO UN AGENTE DEL
INFINITO, LLEVA A LOS
MAS APARTADOS HABI-
TANTES DEL PAIS LA
ENHORABUENA DE LA
NATURALEZA EN EL
CENTENARIO DE LA PATRIA

DEL MISMO
MODO:
Nuestros catálogos - agentes de pro-
greso - llevan á los más lejanos hogares la última
palabra del ARTE y del CONFORT.

PIDANSE NUESTROS CATALOGOS DE MUEBLES
THOMPSON C. Pellegrini, 380
Buenos Aires.

SEEING HALLEY'S COMET: THE CHANCE OF A LIFETIME

20.7 *Halley's Comet, 1910. A photograph taken at the Observatorio Naçional Argentino. (Courtesy, Ruth S. Freitag, Library of Congress.)*

20.8 *Halley's Comet in 1910, as photographed in Helwan, Egypt. (Courtesy, NASA.)*

20.10 *(Opposite) A colour-enhanced photograph of Halley's Comet in 1910. (Courtesy, National Optical Astro Observatory/Lowell Observatory.)*

creation, and towards chauvinism and violence and fear. This battle, on which the fate of the world truly rests, can be illustrated even by a mote of dust. Consider the history of a certain speck of matter dancing in the air before you – formed, it may be, from material ejected billions of years ago by a star on the other side of the Milky Way; wandering for eons in the interstellar dark; gathered up into the forming solar nebula some five billion years ago; adhering to other similar grains in a lump of cosmic matter that eventually formed one of trillions of comets, probably in the vicinity of Uranus and Neptune; ejected out to the Oort Cloud; perturbed into the inner solar system a few thousand years ago; spewed off into the cometary tail, still encased in a particle of ice; wandering as a microplanet in interplanetary space; accidentally intercepted by the Earth a few years ago; gently settling through the Earth's atmosphere; and now bobbing in the air, illuminated by a sunbeam, and giving no apparent hint of its epic cosmic voyage.

Now consider another mote of dust, with a very similar history, but it is a little larger: it enters the Earth's atmosphere and burns up, ionizing the surrounding air, while itself being reduced to molecules and atoms. That bit of matter makes a meteor trail, glowing brightly for a moment, perhaps to the accompaniment of expressions of delight and surprise from the onlookers below. But if you think hard enough, you can find a use even for a shooting star. The trail of ions that a meteor briefly leaves will reflect very high frequency (VHF) radio waves. Since at any given moment there are enormous numbers of meteor trails in the atmosphere – most too faint to be seen with the naked eye – the meteor trails collectively provide a kind of reflecting surface surrounding the Earth, by which radio waves of the proper frequencies can be relayed. Since the average duration of an individual ion trail is less than a second, the message must be sent very fast. This has led to an entire field of technology called Meteor Burst Communications.

And why would anyone go to such lengths, when there are perfectly adequate means of communication at hand? Because communications satellites would be among the early casualties in a nuclear war. Meteor Burst Communications have been developed so that a nuclear war can be fought. The comets have been enlisted. Finally, they have been found to have practical value. But they have not been singled out. The entire worldwide enterprise of human knowledge is being drawn upon for similar obscene services. Something approaching half the scientists on Earth work for the various national military establishments.

Every encounter of Halley's Comet has provided a welcome

opportunity for the expression of bottled-up hopes and fears. Almost ritually, it has been a time to pray. So for this apparition of Halley's Comet, we offer a prayer of our own: we live on a fragile planet, whose thoughtful preservation is essential if our children are to have a future. We are only custodians, for a moment, of a world that is itself only a mote of dust in a universe incomprehensibly vast and old. May we, therefore, learn to act first for the species and the planet.

Then, for innumerable nights to come, there will be humans, or the descendants of humans, to witness its grandeur each time a comet graces the skies of Earth.

Individual comets approach the Sun, flicker a few hundred times, and die like moths around a flame. But a vast repository of them waits at the threshold of the solar system. When the present configuration of continents is unrecognizably altered, when the Earth, engulfed by the expanding Sun, becomes a red giant star, when, in its dotage, the Sun feebly illuminates the charred remains of the mantle of the Earth – then, even then, the skies will still be brightened by the exuberant passage of young comets, newly-arrived from the interstellar dark. When the rest of the solar system is dead, and the descendants of humans long ago emigrated or extinct, the comets will still be soaring.

20.9 *The positions of the tail of Halley's Comet during the 1910 apparition. A drawing by Lucien Rudaux for the French magazine L'Illustration. (Courtesy, Ruth S. Freitag, Library of Congress.)*

In addition to work listed in the Bibliography, further information about comets can be obtained from:

The Planetary Society
65 North Catalina Avenue
Pasadena, CA 91106

The International Halley Watch
Jet Propulsion Laboratory
California Institute of Technology
4800 Oak Grove Drive
Pasadena, CA 91109

For amateur astronomers:

The Astronomical Society of the Pacific
1290 24th Avenue
San Francisco, CA 94122

Other resources for those who wish to dig more deeply include local planetariums and popular science publications such as *Sky and Telescope,* or *Astronomy.*

BIBLIOGRAPHY

BOOKS ON COMETS MAINLY FOR GENERAL AUDIENCES

Brandt, John C., ed. *Comets: Readings from Scientific American.* W. H. Freeman, San Francisco, 1981.

Brandt, John C. *Introduction to Comets.* Cambridge University Press, Cambridge, 1981.

Calder, Nigel. *The Comet is Coming!* Penguin Books, London, 1982.

Chapman, Robert D., and John C. Brandt. *The Comet Book: A Guide for the Return of Halley's Comet.* Jones and Bartlett, Boston, 1984.

Comets: Career Oriented Modules to Explore to Pictures in Science, National Science Teachers Association, 1984.

Dahlquist, Raf, and Theresa Dahlquist. *Mr Halley and his Comet.* Polestar, Canoga Park, Ca., 1982. A charmingly illustrated children's book in rhyme.

Flaste, Richard, Holcomb Noble, Walter Sullivan, and John Noble Wilford. *The New York Times Guide to the Return of Halley's Comet.* Times Books, New York, 1985.

'Halley's Comet'. *The Planetary Report,* 5, (3) May/June, 1985.

Halley Watch Amateur Observers' Manual for Scientific Comet Studies, Enslow Publishers, 1983.

Moore, Patrick. *Guide to Comets.* Lutterworth Press, Guildford, 1977.

Moore, Patrick, and J. Mason. *The Return of Halley's Comet.* W. Stephens, Cambridge, 1984.

Mumford, George. *Everyone's Complete Guide to Halley's Comet.* Sky Publishing Co., Cambridge, 1985.

Pasachoff, Jay M., and Donald H. Menzel. *A Field Guide to the Stars and Planets.* Chapter 11, Houghton Mifflin Company, Boston, 1983.

Rahe, Jurgen, Bertram Donn, and Karl Würm. *Atlas of Cometary Forms: Structures Near the Nucleus.* NASA Special Publication 198. US Government Printing Office, Washington, 1969.

Reddy, F. *Once in a Lifetime: Your Guide to Halley's Comet.* Astromedia, Milwaukee, 1985.

Richter, Nikolaus Benjamin. *The Nature of Comets.* Methuen, London, 1963.

The Royal Institution Library of Science: Astronomy. Vols. 1 and 2, Bernard Lovell, ed. Elsevier, Barking, 1970. A compilation of Friday Evening Discourses on astronomy at the Royal Institution, London, from the middle nineteenth century, including several interesting early talks on meteors and comets.

Seargent, David A. *Comets: Vagabonds of Space.* Doubleday, New York, 1982.

Stasiuk, Garry, and Dwight Gruber. *The Comet Handbook.* Stasiuk Enterprises, 1984.

Wilkening, L., ed. *Comets.* University of Arizona Press, Tucson, 1982. While technical, this collection of review papers is the single most comprehensive reference work on all aspects of comets now in existence.

REPRESENTATIVE TECHNICAL PAPERS AND BOOKS ARRANGED BY CHAPTERS

Chapter 2

Dreyer, J. L. E. *A History of Astronomy from Thales to Kepler.* Cambridge University Press, Cambridge, 1906.

Hasegawa, Ichiro. 'Catalogue of Ancient and Naked-Eye Comets'. *Vistas in Astronomy 24*, 59, 1980.

Hellman, C. Doris. *The Comet of 1577: Its Place in the History of Astronomy.* Columbia University Press, New York, 1944.

Lagercrantz, Sture. 'Traditional Beliefs in Africa Concerning Meteors, Comets, and Shooting Stars'. In *Festschrift fur Ad. E. Jensen.* Klaus Renner Verlag, Munich, 1964.

Leon-Portilla, Miguel, ed. *The Broken Spears: The Aztec Account of the Conquest of Mexico.* Beacon Press, Boston, 1962.

Sarton, George. *A History of Science,* Vol. 1, Harvard University Press, Boston, 1952.

Stein, J., S.J., 'Calixte III et la comète de Halley'. *Specola Astronomica Vaticana II.* Tipografia Poliglotta Vaticana, Rome, 1909.

Thorndike, Lynn. *A History of Magic and Experimental Science.* Vol. 4, Columbia University Press, New York, 1934.

Wen wu. 'Ma Wang Tui po shu "T'ien wen ch'i hsiang tsa chan" nei jung chien shu' and 'Ma Wang Tui Han ts'ao po shu chung to hui hsing t'u', Wen wu ch'u pan she, Beijing, vol. 2, pp. 1-9, 1978.

Yoke, Ho Peng. 'Ancient and Mediaeval Observations of Comets and Novae in Chinese Sources'. *Vistas in Astronomy.*

Ze-zong, Xi. 'The Cometary Atlas in the Silk Book of the Han Tomb at Mawangdui'. *Chinese Astronomy and Astrophysics 8,* 1, 1984.

Chapter 3

Barker, Thomas. *Of the Discoveries Concerning Comets.* Whiston and White, London, 1757.

Eddington, Arthur Stanley. 'Halley's Observations on Halley's Comet, 1682'. *Nature 83,* 373, 1910.

Freitag, Ruth S. *Halley's Comet: A Bibliography.* Library of Congress, Washington, DC, 1984.

MacPike, Eugene, ed. *Correspondence and Papers of Edmond Halley,* Clarendon Press, Oxford, 1932.

Ronan, Colin. *Edmond Halley: Genius in Eclipse.* Macdonald, London, 1969.

Stephenson, F.R., K.K.C. Yao, and H. Hunger. 'Records of Halley's Comet on Babylonian Tablets'. *Nature 314,* 587, April 18, 1985.

Walter, David L. 'Medallic Memorials of the Great Comets and the Popular Superstitions Connected with Their Appearance'. Scott Stamp and Coin Company, New York, 1893.

Westfall, Richard S. *Never at Rest: A Biography of Isaac Newton.* Cambridge University Press, Cambridge, 1980.

White, Andrew Dickson. 'A History of the Doctrine of Comets'. *Papers of the American Historical Association, 2, (2)* G.P. Putnam, New York, 1887.

Chapter 4

Lalande, J J. 'Madame Nicole-Reine Etable de la Briere Lepaute'. *Astronomical Bibliography with a History of Astronomy between 1781 and 1802*. Paris, 1803.

Paulsen, Friedrich. *Immanuel Kant: His Life and Doctrine, 1899*. English Translation by J. E. Creighton and Albert Lefevre, Frederick Ungar, New York, 1963.

Wright, Thomas. *An Original Theory of the Universe*. (Original Printing 1750). Macdonald, London, 1971. This edition has a very useful introduction both to Thomas Wright and his book written by Michael A. Hoskin.

Chapter 6

Feynmann, Richard P., Robert B. Leighton, and Matthew Sands. *The Feynmann Lectures on Physics*. Addison-Wesley Publishing Co., Reading, Massachusetts, 1963. The discussion of the physics of ice in Chapter 6 is partly based upon these remarkable lectures.

Hallett, John. 'How Snow Crystals Grow'. *American Scientist 72*, 582, November/December, 1984.

Patterson, W. S. B. *The Physics of Glaciers*. Pergamon Press, Oxford, 1969.

Whipple, F. L. *Astronomical Journal 11*, 375, 1950.

Whipple, F. L. *Astronomical Journal 113*, 464, 1951.

Chapter 7

Fanale, F., and James Salvail. 'An Idealized Short-Period Comet Model: Surface Insolation, H_2O Flux, Dust Flux, and Mantle Evolution', *Icarus*.

Hughes, David W. 'Cometary Outbursts, A Brief Survey'. *Quarterly Journal of the Royal Astronomical Society 16*, 410, 1975.

Chapter 8

Khare, B. N., and Carl Sagan. 'Experimental Interstellar Organic Chemistry: Preliminary Findings'. In *Molecules in the Galactic Environment*. M. A. Gordon and L. E. Snyder (eds.), John Wiley and Sons, New York, 1973.

Metz, Jerred. *Halley's Comet, 1910: Fire in the Sky*. Singing Bone Press, St. Louis, Mo., 1985.

Mitchell, G. F., S. S. Prasad, and W. T. Huntress. 'Chemical Model Calculations of C_2, C_3, CH, CN, OH, and NH_2 Abundances in Cometary Comas'. *The Astrophysical Journal 244*, 1087, 1981.

Swings, P. 'Le Spectre de la Comète d'Encke 1947 i'. *Annales d'Astrophysique 11*, 1, 1948.

Wood, John, and Sherwood Chang, eds. *The Cosmic History of the Biogenic Elements and Compounds*. NASA Special Publication 476, 1985.

Chapter 9

Barnard, E. E. 'On the Anomalous Tails of Comets'. *The Astrophysical Journal 22*, 249, 1905.

Biermann, L. 'Solar Corpuscular Radiation and the Interplanetary Gas'. *Observatory 77*, 109, 1957.

Biermann, L., and R. Lüst. 'Comets: Structure and Dynamics of Tails'. Chapter 18 of *The Solar System*, vol. 4, B. M. Middlehurst and G. P. Kuiper (eds.), University of Chicago Press, Chicago, 1953.

Henry, George E. 'Radiation Pressure'. *Scientific American 196*, 99-108, 1957.

Van Allen, James A. 'Interplanetary Particles and Fields'. In *The Solar System*, A Scientific American book, W. H. Freeman and Company, San Francisco.

Chapter 10

D'Alviella, Goblet, Count. *The Migration of Symbols*. University Books, New York, 1956.

Donnelly, Ignatius. *Ragnarok: The Age of Fire and Gravel*. University Books, New York, 1970 (original publication, 1883).

Goldsmith, Donald, ed. *Scientists Confront Velikovsky: Evidence Against Velikovsky's Theory of 'Worlds in Collision'*. W. W. Norton, New York, 1977.

Nuttall, Zelia. 'The Fundamental Principles of Old and New World Civilizations: A Comparative Research Based on a Study of the Ancient Mexican Religious, Sociological and Calendrical Systems'. *Archaeological and Ethnological Papers of the Peabody Museum*. Harvard University, vol. 2, 1901.

Wilson, Thomas. *The Swastika: The Earliest Known Symbol, and its Migrations; With Observations on the Migration of Certain Industries in Prehistoric Times*. Smithsonian Institution, Washington, DC, 1896.

Chapter 11

Chebotarev, G. A. 'On the Dynamical Limits of the Solar System'. *Soviet Astronomy – AJ 8*, 787, 1965.

Oort, J. H. 'The Structure of the Cloud of Comets Surrounding the Solar System, and a Hypothesis Concerning Its Origin'. *Bulletin of the Astronomical Institutes of the Netherlands 11*, 91, 1950.

Oort, J. H. 'Empirical Data on the Origin of Comets'. Chapter 20 in G. P. Kuiper and B. M. Middlehurst (eds.), *The Solar System*, vol. 4., University of Chicago Press, Chicago, 1953.

Öpik, E. 'Note on Stellar Perturbations of Nearly Parabolic Orbits', *Proceedings of the American Academy of Arts and Sciences 67*, 169, 1932.

Russell, Henry Norris. *The Solar System and Its Origin*. Macmillan, New York, 1935.

Van Woerkom, A. J. J. 'On the Origin of Comets'. *Bulletin of the Astronomical Institutes of the Netherlands 10*, 445, 1948.

Chapter 12

Biermann, L., and K. W. Michel. 'On the Origin of Cometary Nuclei in the Presolar Nebula'. *The Moon and the Planets 18*, 447, 1978.

Fernandez, Julio A. 'Mass Removed by the Outer Planets in the Early Solar System'. *Icarus 34*, 173, 1978.

Goldreich, P., and W. R. Ward. 'The Formation of Planetesimals'. *The Astrophysical Journal 183*, 1051, 1973.

Helmholtz, H. 'On the Origin of the Planetary System'. Lecture delivered by H. Helmholtz in Heidelberg and Cologne, 1871. Published in *Popular Articles on Scientific Subjects*, by H. Helmholtz. D. Appleton and Company, New York, 1881.

Safronov, V. S. 'Evolution of the Protoplanetary Cloud and the Formation of the Earth and the Planets'. 1972 (Russian Citation). English Language version published by Israel Program for Scientific Translations, Jerusalem, 1972.

Wetherill, George W. 'Evolution of the Earth's Planetesimal Swarm Subsequent to the Formation of the Earth and Moon'. *Proceedings of the Eighth Lunar Science Conference*, p. 1, 1977.

Chapter 13

Ball, Robert. *The Story of the Heavens*. Revised edition. Cassell and Company, London, 1900.

Keesing's Contemporary Archives. May 21-28, 17425-17429, 1960.

Von Humboldt, Alexander. 'Events of the Night of Eleventh November 1799'. *Personal Narrative of Travels to the Equinoctial Regions of America During the Years 1799 to 1804*, vol. 1. George Bell and Son, London, 1889.

Chapter 14

Gold, Thomas, and Steven Soter. 'Cometary Impact and the Magnetization of the Moon'. *Planetary and Space Sciences 24*, 45, 1976.

Kerr, Richard A. 'Could an Asteroid be a Comet in Disguise?' *Science 227*, Feb. 22. 1985.

Marsden, B. G. 'The Sungrazing Comet Group'. *Astronomical Journal 72*, 1170, 1967.

Michels, D. J., N. R. Sheeley, R. A. Howard, and M. J. Koomen. 'Observations of a Comet on Collision Course with the Sun'. *Science 25*, 1097, 1982.

Öpik, Ernst. 'The Stray Bodies in the Solar System. Part 1. The Survival of Cometary Nuclei in the Asteroids'. *Advances in Astronomy and Astrophysics 2*, 219, 1963.

Öpik, Ernst. 'The Stray Bodies in the Solar System. Part 2. The Cometary Origin of Meteorites'. *Advances in Astronomy and Astrophysics 4*, 301, 1966.

Shoemaker, Eugene M., and Ruth F. Wolfe. 'Cratering Timescales for the Galilean Satellites'. Chapter 10 in *The Satellites of Jupiter*, David Morrison, ed. University of Arizona Press, Tucson, 1982.

Shulman, L. M. 'The Evolution of Cometary Nuclei'. In G. A.

Chebotarev et al., eds. *The Motion, Evolution of Orbits, and Origin of Comets.* p. 271. D. Reidel, Holland, 1972.

Wetherill, George W. 'Occurrence of Giant Impacts During the Growth of the Terrestrial Planets'. *Science 228*, 877, 17 May 1985.

Chapter 15

Alvarez, Luis W., Walter Alvarez, Frank Asaro, and Helen V. Michel. 'Extraterrestrial Cause for the Cretaceous-Tertiary Extinction'. *Science 208*, 1095, 1980.

Gould, Steven Jay. 'Sex, Drugs, Disaster, and the Extinction of the Dinosaurs'. *Discover*, March 1984, p. 67.

Hills, J. G. 'Comet Showers and the Steady-state Infall of Comets from the Oort Cloud'. *Astronomical Journal 89*, 1730, 1981.

National Museum of Natural Sciences and National Research Council of Canada, Syllogeus Series Number 12. 'Cretaceous-Tertiary Extinctions and Possible Terrestrial and Extraterrestrial Causes: Proceedings of a Workshop Held in Ottawa, Canada, 16, 17 November, 1976'. National Museums of Canada, Ottawa, Canada, March, 1977.

Officer, Charles B., and Charles L. Drake. 'Terminal Cretaceous Environmental Events'. *Science 227*, 1161, 1985.

Pollack, James B., Owen B. Toon, Thomas P. Ackerman, Christopher P. McKay, and Richard P. Turco. 'Environmental Effects of an Impact-generated Dust Cloud: Implications for the Cretaceous-Tertiary Extinctions'. *Science 219*, 287, 1983.

Sepkoski, John. 'Mass Extinctions in the Phanerozoic Oceans: A Review'. *Geological Society of America, Special Paper 190*, 283, 1982.

Shoemaker, Eugene M. 'Asteroid and Comet Bombardment of the Earth'. *Annual Review of Earth and Planetary Sciences 11*, 461, 1983.

Steel, Rodney, and Anthony Harvey (eds.). *The Encyclopaedia of Prehistoric Life.* McGraw-Hill, New York, 1979.

'The Fossil Record and Evolution: Readings from *Scientific American*'. W. H. Freeman and Company, San Francisco, 1982.

Urey, Harold C. 'Cometary Collisions and Geological Periods'. *Nature 242*, 32, 1973.

Chapter 16

Alvarez, Walter, Frank Asaro, Helen V. Michel, and Luis W. Alvarez. 'Iridium Anomaly Approximately Synchronous with Terminal Eocene Extinctions'. *Science 216*, 886, 1982.

'A Talk with Eugene Shoemaker'. (Interview by Charlene Anderson), *The Planetary Report 5* (1) 7, January/February, 1985.

Davis, Marc, Piet Hut, and Richard A. Muller. 'Extinction of Species by Periodic Comet Showers'. *Nature 308*, 715, 1984.

Gould, Steven Jay. 'Continuity'. *Natural History*, April 1984, p.4.

Gould, Steven Jay. 'The Cosmic Dance of Siva'. *Natural History*, August 1984, p.14.

Hills, J. G. 'Dynamical Constraints on the Mass and Perihelion Distance of Nemesis and the Stability of Its Orbit'. *Nature 311*, 636, 1984.

Hoffman, Antoni. 'Patterns of Family Extinction Depend on Definition and Geological Timescale'. *Nature 315*, 659, 1985.

Rampino, Michael, and Richard Stothers. 'Geological Rhythms and Cometary Impacts'. *Science 226*, 1427, 1984.

Rampino, Michael, and Richard Stothers. 'Terrestrial Mass Extinctions, Cometary Impacts and the Sun's Motion Perpendicular to the Galactic Plane'. *Nature 308*, 709, 1984.

Raup, David M., and J. John Sepkoski, Jr. 'Periodicity of Extinctions in the Geologic Past'. *Proceedings of the National Academy of Sciences of the USA 81*, 801, 1984.

Schwartz, Richard D., and Philip B. James. 'Periodic Mass Extinctions and the Sun's Oscillations about the Galactic Plane'. *Nature 308*, 712, 1984.

Thaddeus, Patrick, and Gary A. Chanan. 'Cometary Impacts, Molecular Clouds, and the Motion of the Sun Perpendicular to the Galactic Plane'. *Nature 314*, 73, 1985.

Whitmire, Daniel, and Albert A. Jackson. 'Are Periodic Mass Extinctions Driven by a Distant Solar Companion?' *Nature 308*, 713, 1984.

Chapter 17

Bar-Nun, A., A. Lazcano-Araujo, and J. Oro. 'Could Life Have Evolved in Cometary Nuclei?' *Origins of Life 11*, 387, 1981.

Forster, T. *Atmospheric Causes of Epidemic Diseases*. London, 1829.

Hobbs, R. W., and J. M. Hollis. 'Probing the Presently Tenuous Link between Comets and the Origin of Life'. *Origins of Life 12*, 125, 1982.

Hoyle, Fred. 'Comets – A Matter of Life and Death'. *Vistas in Astronomy 24*, 1980.

Hoyle, Fred, and Chandra Wickramasinghe. *Diseases from Space*. Sphere Books, London, 1981.

Irvine, W. M., S. B. Leschine, and F. P. Schloerb. 'Thermal History, Chemical Composition and Relationship of Comets to the Origin of Life'. *Nature 283*, 748, 1980.

Oro, J. 'Comets and the Formation of Biochemical Compounds on the Primitive Earth'. *Nature 190*, 389, 1961.

Shoemaker, Eugene M. 'Asteroid and Comet Bombardment of the Earth'. *Annual Review of Earth and Planetary Sciences 11*, 461, 1983.

Wallis, Max. 'Radiogenic Melting of Primordial Comet Interiors'. *Nature 284*, 431, 1980.

Chapter 18

Gay, Peter. *Education of the Senses* (Vol. 1 of *The Bourgeois Experience: Victoria to Freud*). Oxford University Press, Oxford, 1984.

'Halley's Comet from a Balloon'. *Aeronautics 6*, 204, June, 1910.

Morrison, David, et al. *Planetary Exploration Through the Year 2000: A Core Program*. National Aeronautics and Space Administration, US Government Printing Office, Washington, DC, 1983.

Neugebauer, M., et al., eds. *Space Missions to Comets*. NASA Conference Publication 2089, 1979.

Newburn, Ray L., and Jurgen Rahe. 'The International Halley Watch'. *Journal of the British Interplanetary Society 37*, 28, 1984.

Newsletters of the International Halley Watch. Periodicals produced by the NASA Jet Propulsion Laboratory, Publications 410, since August 1, 1982.

Planetary Society Fact Sheets on Halley Comet Spacecraft, available from The Planetary Society, 65 North Catalina Avenue, Pasadena, CA 91106.

'Report of the Comet Rendezvous Science Working Group'. NASA Technical Memorandum 87564, 1985.

Sagdeev, R. Z., et al. 'Cometary Probe of the Venera-Halley Mission'. *Advances in Space Research 2*, 83, 1983.

Wilford, John Noble. 'US and Soviet Cooperating on Collection of Comet Dust'. *New York Times*, December 21, 1984.

Yeomans, Donald K., and John C. Brandt. 'The Comet Giacobini-Zinner Handbook: An Observer's Guide'. NASA/JPL Publication 400-254, 1985.

Harwit, Martin, and E. E. Salpeter. 'Radiation from Comets near Neutron Stars'. *The Astrophysical Journal 186*, L37, 1973.

MIT Student Project in Systems Engineering, 'Project Icarus'. MIT Press, Cambridge, Massachusetts, 1968.

Chapter 19

Dyson, Freeman. 'The World, the Flesh, and the Devil'. Third J. D. Bernal Lecture, delivered at Birkbeck College, London. Reprinted in C. Sagan, ed. *Communication with Extraterrestrial Intelligence (CETI)*, MIT Press, Cambridge, Mass., 1973.

Gaffey, Michael J., and Thomas B. McCord. 'Mining Outer Space'. *Technology Review 79* (7), 51, June 1977.

Chapter 20

Arkin, William M., and Richard W. Fieldhouse. *Nuclear Battlefields: Global Links in the Arms Race*. Ballinger, Cambridge, Mass., 1985.

Cruikshank, D. P., W. K. Hartmann, and D. J. Tholen. 'Colour, Albedo, and Nucleus Size of Halley's Comet'. *Nature 315*, 122, 1985.

Newburn, Ray L., and Donald K. Yeomans. 'Halley's Comet'. *Annual Review of Earth and Planetary Science 10*, 297, 1982.

APPENDIX

COMETARY ORBITS AND METEOR SHOWERS

ORBITS OF SELECTED LONG-PERIOD COMETS

Comet	Name	Perihelion Distance (AU)
1811 I	Great Comet	1.035
1844 III	Great Comet	0.251
1858 VI	Donati	0.578
1861 II	Great Comet	0.822
1881 III	Great Comet	0.735
1882 II	Great Comet	0.008
1908 III	Morehouse	0.945
1910 I	Great Comet	0.129
1937 IV	Whipple	1.734
1943 I	Whipple-Fedtke-Tevzadze	1.354
1957 III	Arend-Roland	0.316
1957 V	Mrkos	0.355
1962 VIII	Humason	2.133
1965 VIII	Ikeya-Seki	0.008
1969 IX	Tago-Sato-Kosaka	0.473
1973 XII	Kohoutek	0.142
1976 VI	West	3.277
1980b	Bowell	3.364

Note that perihelion passage for these long-period comets varies from very close to the Sun (0.008 AU) to the middle of the asteroid belt, between the orbits of Jupiter and Mars (1.4 to 5 AU). There are probably long-period comets with more distant perihelia that are unknown to observers on Earth.

Data from B. Marsden and E. Roemer, *Comets* (L. Wilkening, ed.), University of Arizona Press, 1982.

ORBITS OF SELECTED SHORT-PERIOD COMETS

Periodic Comet	Perihelion Passage	Inclination in degrees	Perihelion Distance (AU)	Eccentricity	Period (years)
Oterma	1983 June	1.94	5.4709	0.2430	19.4
Crommelin	1984 Feb.	29.10	0.7345	0.9192	27.4
Giacobini-Zinner	1985 Sept.	31.88	1.0282	0.7076	6.59
Halley	1986 Feb.	162.23	0.5871	0.9673	76.0
Whipple	1986 June	9.94	3.0775	0.2606	8.49
Encke	1987 July	11.93	0.3317	0.8499	3.29
Borelly	1987 Dec.	30.32	1.3567	0.6242	6.86
Tempel 2	1988 Sept.	12.43	1.3834	0.5444	5.29
Tempel 1	1989 Jan.	10.54	1.4967	0.5197	5.50
d'Arrest	1989 Feb.	19.43	1.2921	0.6246	6.39
Perrine-Mrkos	1989 Mar.	17.82	1.2977	0.6378	6.78
Tempel-Swift	1989 Apr.	13.17	1.5884	0.5391	6.40
Schwassmann-Wachmann 1	1989 Oct.	9.37	5.7718	0.0447	14.9
Kopff	1990 Jan.	4.72	1.5851	0.5430	6.46
Tuttle-Giacobini-Kresak	1990 Feb.	9.23	1.0680	0.6557	5.46
Honda-Mrkos-Pajdusakova	1990 Sept.	4.23	0.5412	0.8219	5.30
Wild 2	1990 Dec.	3.25	1.5779	0.5410	6.37
Arend-Rigaux	1991 Oct.	17.89	1.4378	0.6001	6.82
Faye	1991 Nov.	9.09	1.5934	0.5782	7.34

The inclinations of the comets shown here vary from less than 2° (almost exactly in the ecliptic plane which includes the Earth and the planets), to 162° for Halley's Comet. Since a 90° inclination would mean a comet orbiting the Sun perpendicular to the ecliptic, this means a comet whose orbit is inclined some (180-162=) 18° from the ecliptic, but travelling in the opposite sense to the direction in which the planets revolve about the Sun. The eccentricity of the cometary orbits shown varies from about 0.04 (very close to a circle) to 0.97 for Halley's Comet, which is extremely elongated.

Data from Marsden and Roemer (op. cit.).

METEOR SHOWERS VISIBLE TO THE NAKED EYE, LATE TWENTIETH CENTURY

Shower	Shower Maximum Date	Single Observer Hourly Rate	Speed of Encounter with Earth in kilometres/sec.	Normal Duration to 1/4 Strength of Max., in days	Associated Comet
Quadrantids	Jan. 3	40	41	1.1	?
Lyrids	Apr. 22	15	48	2	1861 I
Eta Aquarids	May 4	20	65	3	Halley
S. Delta Aquarids	July 28	20	41	7	?
Perseids	Aug. 12	50	60	4.6	1862 III
Orionids	Oct. 21	25	66	2	Halley
S. Taurids	Nov. 3	15	28	—	Encke
Leonids	Nov. 17	15	71	—	1866 I
Geminids	Dec. 14	50	35	2.6	(1983 TB)?
Ursids	Dec. 22	15	34	2	Tuttle

Data based on table by P. M. Millman in *Observer's Handbook* of the Royal Astronomical Society of Canada (1985).

INDEX